DEFLATING INFORMATION:
FROM SCIENCE STUDIES TO DOCUMENTATION

Bernd Frohmann

Deflating Information
From Science Studies to Documentation

UNIVERSITY OF TORONTO PRESS
Toronto Buffalo London

© University of Toronto Press Incorporated 2004
Toronto Buffalo London
Printed in Canada

ISBN 0-8020-8839-2

Printed on acid-free paper

National Library of Canada Cataloguing in Publication

Frohmann, Bernd Peter, 1946-
Deflating information : from science studies to documentation /
Bernd Frohmann.

Includes bibliographical references and index.
ISBN 0-8020-8839-2

1. Documentation.   2. Information science.   3. Science – Social aspects.
4. Technical writing.   I. Title.

Q175.F76 2004   306.4'2   C2004-901810-8

This book has been published with the help of a grant from the Canadian
Federation for the Humanities and Social Sciences, through the Aid to
Scholarly Publications Programme, using funds provided by the Social
Sciences and Humanities Research Council of Canada.

University of Toronto Press acknowledges the financial assistance to its
publishing program of the Canada Council for the Arts and the Ontario
Arts Council.

University of Toronto Press acknowledges the financial support for its
publishing activities of the Government of Canada through the Book
Publishing Industry Development Program (BPIDP).

For my parents

Contents

Acknowledgments

Since this book is primarily a synthesis of the work of others, my major debt is to them and is repaid, however inadequately, in notes and references. Like many of the best-laid research plans, my intended journey revealed new and irresistible trails along the way. A plan to explore how information and 'information users' became objects of theoretical speculation, or how they emerged as entities about which a specific kind of knowledge may be had, led me directly to information in its most robust form: scientific information. The exploration of what I first believed was but a preparatory topic grew to a book-length manuscript and occupied me throughout my sabbatical leave of 1997–8. I would like to thank colleagues at two universities who generously sponsored me as a visiting scholar during that time: Fran Miksa at the Graduate School of Library and Information Science, University of Texas at Austin, and John V. Richardson, Jr, at the Department of Information Studies, University of California at Los Angeles (UCLA). It is not possible to list all the people at these institutions who so generously responded to my presentations of ideas that found their way into this book, but I am grateful to them all. I also thank Sandra Harding, who generously admitted an unknown researcher into her invaluable seminar on science studies at UCLA. I have reason to be grateful many times over for the hospitality of hearth and spirit extended to me by Michael Buckland of the School of Information Management and Systems, University of California, Berkeley, and for his kind invitations to permit an airing of some of the ideas of this book in his stimulating Friday seminars. Julie Hallmark of UT Austin, and her daughters Anne and Margaret Bichteler, contributed immeasurably by their warm hospitality and kind generosity in creating spaces where work was always a pleasure. A special thank-you to Susan Ehrlich of York University, Toronto, for our conversations about some of the issues discussed in this book. Students in my graduate course in the winter term of 2003 helped clarify my ideas, and their contribution is evident here; I thank Louis Beaubien, Fiona Coll, George Goodall, Kate Kerr, Jon van der Veen, and Peter West. Siobhan Stevenson and Brendan Luyt worked for me as research assistants while pursuing their doctoral studies, and I thank them for their labours.

This book could not have been produced without the technical assistance of John Fracasso, Coordinator, Computer Services, and Charles McClellan, Network Specialist, both of the Faculty of Information and Media Studies at The University of Western Ontario. Their expertise, extraordinary patience, and creative solutions to problems cannot be appreciated enough.

I gratefully acknowledge the generous support provided for the research presented here by the award of an individual research grant for the period 1996–9 by the Social Sciences and Humanities Research Council of Canada. The manuscript also benefitted from the careful attention of reviewers of the Council's Aid to Scholarly Publications Programme and for the University of Toronto Press; I am indebted to them. I am also grateful for the expert copy-editing work of Allyson N. May of the University of Toronto Press, who improved infelicities of style in many places.

Finally, I owe much to Marilyn Randall, whose daily patience and generosity lighten the burdens of pursuing two academic careers under one roof.

DEFLATING INFORMATION

Introduction:
From Information to Documentation

The Paradox of Scientific Documentation

The natural sciences are undoubtedly the most studied disciplines in research on scholarly communication and information use.[1] Yet since J.D. Bernal's 1948 report for the Royal Society's Scientific Information Conference on the information needs of scientists, plausibly the founding text of this research area, scientific documents have occupied a precarious position in studies of science information systems. On the one hand, they have been called 'the most important source and medium of scientific information' (Mikhailov, Chernyi, and Giliarevski 1984, 198). They have even been identified with science itself: 'Scientific documents are a form of science. Without them, science cannot exist' (147).[2] Yet their exponential growth, coinciding with vast infusions of funding for scientific research and development after the Second World War, was seen by many observers not as a sign of the robust good health of the science information system, but rather as cause to fear its imminent collapse and ruin. In a typical assessment of 'the chaos of scientific publication,' published almost a decade before his Royal Society study, Bernal lamented that 'the burden of this vast mass of publication is in itself a great handicap to scientific research' (1939, 119, 118). The tremors of our contemporary 'information anxiety' (Wurman 1989) were felt well over half a century ago in the field of scientific documentation.

Worries about the uncontrolled proliferation of science's formal literature – laments about the eternal burial, under the crushing weight of masses of documents, of crucial findings and the important conclusions they might have generated in spite of the heroic but ultimately doomed attempts of the system's information retrieval designers to keep abreast of its galloping rate of publication – were not the worst of it. Not only was there simply too much of the stuff, but most of it turned out to be useless. Reinforcing an observation of many writing before him, Robert Hayes concluded that 'natural scientists are focused on the acquisition of new data rather than the analysis of existing records. For them, the records of the past are peripheral to research' (1992, 6). By the time it gets into the hands of research scientists, the formal literature is already outdated. It therefore came as no surprise that, after quoting a Nobel laureate on the possibility of a paper's going unrecognized, even to the extent of nobody reading it, Robert K. Merton remarked: 'Studies of reading practices of scientists indicate that the suggested possibility – "Nobody reads it" – is something less than sheer hyperbole. It has been found, for example, that only about half of one percent of the articles published in journals of chemistry are read by any one chemist. And much the same pattern has been found to hold in psychology' (1973b, 448).

Sharpening the point by quantitative methods, Derek de Solla Price's famous eponymous index (1970; repr. in Price 1986) gave mathematical expression to the recurrent conclusion of studies of science information systems, that this archive of 'the records of the past' is consulted with such a low frequency that other, and most typically informal, channels of scientific communication are the ones most critical to research work. 'There is plenty of evidence to suggest,' Blaise Cronin remarks, 'that most scientists are content to operate, and participate in, a two-tier system, with informal networks being used to channel preliminary notifications of research findings, and the formal system being used as an archive of findings which have been vetted and approved by the scientific establishment' (1984, 12). The theme of the 'two-tier system' is echoed by Hayes: 'The role of personal communication in information access appears to be much more important to research in the natural sciences than it is in the humanities and social sciences. For the scientist, speed of communication is crucial since the results obtained by one colleague immediately influence the work of others with similar foci of research. This is reflected in the historic importance of preprints, telephone, and now electronic communication to those sciences' (1992, 6).

The final indignity suffered by science's beleaguered formal literature, in addition to its unbridled propagation and precipitous drop in value as a meaningful research resource, is its stature among the few who actually read it. Bernal's indictment of the literature's quality was scathing: 'The very bulk of scientific publication is itself delusive. It is of very unequal value; a large proportion of it, possibly as much as three-quarters, does not deserve to be published at all, and is only published for economic reasons having nothing to do with the real interests of science' (1939, 118). And in a report that would likely raise few eyebrows in contemporary academic communities, the authors remark that '*all* our respondents thought that the vast majority of papers in the journals which they read were of poor quality or of little significance. One respondent stated that "only 2 per cent of publications are significant and 50 per cent are wrong." This response may be phrased in extreme terms but every member of the department endorsed its general sentiments' (Mulkay and Williams 1971, 74). Bruno Latour and Steve Woolgar sum up the situation by saying, 'Most published papers are never read, the few that are read are worth little, and the remaining 1 or 2 percent are transformed and misrepresented by those who use them' (1986, 252).

Yet paradoxically, the mass of scientific documents also seems to picture *science itself* in its most graphic and unequivocal form. A journal article appears to embody the very substance of science, because it is imagined as containing and conveying a small bit of the information which, together with the information contained in other articles, constitute the field of scientific knowledge of the natural world. All the television broadcasts, speeches, films, CD-ROMs, websites, and magazine and encyclopedia articles designed to assist the lay public in gaining some measure of scientific knowledge are merely pale reflections of the real thing. When it comes to representing nature directly, the formal literature is where the action is. The natural sciences and their formal, peer-reviewed literature have long provided a standard for evaluating the claims of pretenders to the title of any organized, systematic knowledge.[3] By representing research results directly (albeit not immediately), the formal literature communicates scientific information with an immediacy lost in the reworked, edited, sanitized, and often adulterated form of the secondary, tertiary, and popular literatures of science.[4] The journal literature therefore seems to present an image of science itself: if one had a perspicuous representation of the whole field of scientific documentation, one would see all the scientific information there is; one could map the state of scientific knowledge at any particular moment.[5] In other words, the journal literature, by containing scientific information, conveys raw materials for the production of scientific knowledge. It is for this reason that Mikhailov, Chernyi, and Giliarevski call it 'a form of science.' By communicating the facts, observations, methods, and theories that make up scientific knowledge of the natural world, the journal article has enjoyed its position at the core of scientific communication and information transfer since the origins of modern science in the mid-seventeenth century.[6] Today it is selected, acquired, indexed, abstracted, catalogued, compiled in bibliographies, stored in and retrieved from electronic databases and the World Wide Web, and implicated in the many elaborate, costly, and labour-intensive procedures of complex information retrieval systems.

What is the source of this paradox of scientific documentation? How can the most important medium of scientific communication, one thought to be essential to science, indeed, thought itself to be 'a form of science,' be at the same time not only marginal to work at the research front, but also pose grave threats to the effective communication of the very information required for scientific knowledge production? How could a literary form with an ancestry so venerable as to boast the origins of both modern scientific method and the proper literary form for documenting its results become virtually irrelevant at the primary sites of natural-scientific knowledge production?[7]

The Abstract Idea of Information

This book suggests that the paradox arises from a discourse of scientific knowledge that privileges the communication of information at the expense of labour, work, and material practices. The assumption that seeking and communicating information are central to the scientific enterprise both reinforces and is reinforced by an idealization of science that privileges thinking in the service of theory construction. The idea of representation is important here too. Ian Hacking (1983) reminds us that scientific theories are representations. He also argues that the concept of reality is a second-order concept. People make representations of all kinds – 'They paint pictures, imitate the clucking of hens, mould clay, carve statues, and hammer brass' (132) – and once the representations are out there, judgments can be made about their faithfulness, truth, or reality. The idea of reality is parasitic upon representations: it performs the work of judging how well or ill representations depict. A faithful representation depicts truly: it represents reality (or the world). Theories are high-level representations. When they conflict, arguments can arise over which theory represents the world (or reality) truly. Contests between alternative theories spawn philosophical debates of realism versus anti-realism about the question of whether theories represent at all or whether they are merely instruments for thinking: 'As soon as representations began to compete, we had to wonder what is real. Anti-realism makes no sense when only one kind of representation is around. Later it becomes possible' (139). I suggest that the concept of *information* belongs to this nest of concepts and conflicts. Accurate, true, and faithful information is what scientists are imagined as needing to think about the world theoretically. Whether gathered by direct or mediated observation or by experiment, or communicated by documents or colleagues, information is seen as the raw material of scientific thought exercised in theory production. Reliable information about the world is the stuff scientists use to build its theoretical representations.

The main features of this 'thought and theory' idealization of science are: (1) the main work of science is theory construction; (2) a theory is a highly abstract representation of the structure of the natural world; (3) theories are constructed by thought; (4) since thinking is the main work of science, the scientist is primarily a reasoner; (5) scientific thought is the performance of scientific rationality, a disciplined process of reasoning refined over the centuries since its origins in Francis Bacon's attacks on Aristotelian ideas about how knowledge of nature is acquired; and (6) science aims at synthesizing its results in a single theory: a unified representation of the natural world. These six features should not be seen as jointly necessary and sufficient conditions of a logically coherent belief about the nature of science, but rather as tacit assumptions and ways of talking about science that overlap, with the emphasis on one or the other varying from one instance to another. They assign key roles to information seeking, information processing, and the communication of information. Scientists are depicted as gatherers of information, observing, experimenting, reading documents, and communicating with colleagues; as processors of information, applying scientific rationality to the information they gather; as creators of new, higher-level information, drawing conclusions, formulating hypotheses, and generating theoretical propositions; and as synthesisers of information, integrating processed information into a unified, theoretical representation of the natural world. Research problems are generated for studies of scientific information, primary among them the task of finding the most important channel for the communication of the information scientists need for theory production. The idea of information as the abstract object sought, processed, communicated, and synthesized, the raw material of science's theoretical representation of the world and the stuff of scientific thought, is part of the picture of science as thought, theory, and representation.

But this abstract idea of information cuts two ways when used to understand the point and purpose of scientific documentation. It sets the stage for the paradox of the formal literature by simultaneously supporting and undermining its significance for research front work. When information is imagined as the raw material of scientific thought, then by virtue of embodying precisely such information, science's formal literature plays a leading role. But when the same abstract concept of information figures in questions about its most effective vehicles of communication, empirical research shows that the formal literature is not used at science's primary production site: the research front. Under the spell of thought, representation, ideas, conceptual networks, propositions, and conceptions of scientific work as primarily a highly abstract and theoretically oriented cognitive activity, and under the assumption that the pace of contemporary scientific work requires speeds of information transfer impossible to attain through the refereeing systems and print medium of the journal literature, information studies seeks other channels for the communication of scientific information. Informal channels are those most often cited as the most significant; William D. Garvey, for example, observes: 'the journal article mainly serves to provide *new* information to those researchers who were not part of the informal active process by which information contained in articles is disseminated well before (an average of one year) the journal's publication' (1979, 308). Once alternative channels are discovered, the significance of the journal literature is explained in other ways. A familiar explanation, first presented by Merton and his followers, leans on ideas of credit, reward, gift-giving, or the accumulation of intellectual capital.[8]

For information studies, the most important feature of the abstract idea of information, and the assumption underlying the hunt for the principal channel of scientific communication, is the conception of information as a particular kind of substance that can be indifferently conveyed through different channels. This conception has recently come under fire in areas primarily concerned with documentation in its various forms. John Seely Brown and Paul Duguid contrast the social to the 'informational' life of documents by challenging the notion that documents are a 'form of transport onto which information is loaded,' and that they *contain, hold, carry,* and *convey* information' (2000, 184). Such ideas, they argue, blind us to important effects of the material properties of documents, such as their role in creating and maintaining social groups and coordinating social practices: 'documents ... help structure society, enabling social groups to form, develop, and maintain a sense of shared identity. Viewing documents as mere information carriers overlooks this social role' (189). And in his 'Farewell to the Information Age,' Geoffrey Nunberg traces the mid-nineteenth-century emergence of our contemporary 'phenomenon of information' as 'a noble substance ... indifferent to the transformation of its vehicles' (1996, 107), a notion that underwrites, inter alia, the familiar quantification of information which supports its commoditization in the enthusiastic celebrations of the 'information society.' This concept of information is, Nunberg argues, intimately bound up with documents, because we think of information as their *content.* To conceive of information in this way, as an abstract entity only contingently related to its modes of communication, supports the 'information-wants-to-be-free' rhetoric of the digital age: 'In the print world, it was attached to things or contained in them, but now it can be liberated and manipulated as a kind of pure essence: we can break the bottles and have the wine' (107). Nunberg also suggests a connection between this abstract sense of 'information' and the idealist underpinnings of representationalist epistemology when he describes information as 'a kind of intentional substance that is present in the world' (110).

A Wittgensteinian Interlude

Challenges such as these, however, have done little to dislodge the conception of information as a distinct kind of abstract object from its privileged position in studies of science information systems, and in information studies more generally. I would like to pursue briefly the thought that a reason for its persistence has to do with a deeply rooted picture of *meaning,* one probed by Ludwig Wittgenstein in the opening pages of his *Philosophical Investigations* (1958). My intention in introducing it here is to make some connections between it and a closely related picture of information. Wittgenstein's deflationary treatment, or deconstruction, of what he calls 'a picture of the essence of human language' – the 'Augustinian picture' – seems to me to provide us with useful resources for thinking about how documents and information are related, without construing information as a kind of thing or substance, whether imagined as conveyed and communicated by documents, sought by 'information seekers,' or something towards which 'information seeking behaviour' is directed. This brief sketch of the connections between Wittgenstein's approach to meaning and a parallel approach to information allows me to map out a broad framework for a *documentalist* approach to information studies.

The value for my purposes of Wittgenstein's deconstruction of the Augustinian picture follows from three points: (1) the close relation that exists between meaning and information; (2) the level at which his investigation works; and (3) the role of documents in it. The first point recognizes some simple and unexciting connections between the grammar of 'meaning' and the grammar of 'information.' It would make little sense to say that one could be informed by a meaningless document. Although it would be odd to say that one was informed by a single word, it seems less odd to say that if the word were meaningless, there is no chance of being informed by it. Nonetheless, there are examples of single symbols that inform, such as those found on roadway signs. And a meaningless symbol on such a sign will not inform you of anything unless, of course, you have instructed your passengers to consult the legend of roadway signs whenever you encounter one that is meaningless, thereby setting up a meaning for a 'meaningless' sign, at least in the minimal sense of distinguishing it as the kind of sign that informs your passengers they have to consult the legend. So even here, meaning and informing are connected. Nor need we restrict the grammars of meaning and informing to texts, signs, and symbols. Documentalists of the late nineteenth and early twentieth centuries, such as Paul Otlet, admit 'natural objects, artifacts, objects bearing traces of human activity (such as archaeological finds), explanatory models, educational games, and works of art' (Buckland 1997, 805) as *documents*. The paradigm case is perhaps the antelope of 'Madame Documentation,' Suzanne Briet, who allowed that even this creature, once taken to a zoo and made an object of study, counts as a document. Why? As Buckland explains it, these objects 'can be regarded as "documents" if you are *informed* by observation of them' – if they become 'signifying objects' (805; emphasis added). In a zoo, museum, or some other context in which these objects are meaningful, they can be informing. Thus 'meaning,' 'information,' and 'document' have grammatical connections. A document can be informing when it has meaning; without meaning, there can be no information. Since the concepts of meaning, information, and documents are intertwined, a reorientation or displacement of one affects the other two. Wittgenstein's investigation of Augustine's picture of meaning thus has significant implications for rethinking information, documents, and their connections.

The second point speaks to one of the main objectives of this book: *deflating information.* I argue that the idea of information as an abstract object – whether operating as a shadowy and unexamined substantive guiding uses of the word 'information,' as the object of explicit theoretical reflection in information studies, or as the cultural phenomenon Nunberg describes as 'intentional substance present in the world' – directs attention away from investigations of the complex circuits in which documents travel and their equally complex and varied effects.[9] Since the abstract idea of information is deeply lodged, the job of dislodging it requires resources operating at the appropriate depth. Wittgenstein's philosophical investigations not only provide such resources, they also target a concept – an abstract concept of meaning – that is grammatically and conceptually related to an abstract concept of information. The power of his investigations depends upon two things: first, sympathy for his method of 'showing the fly the way out of the fly-bottle' (§309), that is, his way of deconstructing what he claims are conceptual confusions deeply embedded in the grammar of our language, and second, a conviction that the scene of problems in which Wittgenstein works is still our scene today, rather than one narrowly circumscribed by the problems of mind, language, and logic that obsessed Bertrand Russell and Gottlob Frege but from which we have been happily released. Stated in terms of the issues of this book, one needs to find it plausible that there is an abstract idea of information at work in our intellectual culture, and that it works at a level deeper than explicit theorizing, even though its traces are found there.[10] The second and third chapters of this book try to show that such a concept is alive in studies of scientific information and is connected to a 'thought and theory' view of science. The remaining chapters attempt to reveal how recent social studies of science provide resources for reconceptualizing studies of scientific information as studies of scientific documentation. To that extent, I provide support for my claim that an abstract idea of information is still part of at least one corner of our current intellectual scene. As for the broader question of the power of Wittgenstein's treatment of meaning with respect to contemporary issues, one would need to follow, throughout the *Philosophical Investigations* and other works (see especially 'The Brown Book,' in Wittgenstein 1969), his pursuit of the myriad ramifications of the Augustinian picture in our language and thought. This cannot be done here, where my aim is modest: to provide a brief sketch of his treatment of this picture in the opening passages of the *Philosophical Investigations* in order to draw out a way of thinking about documents without resort to an abstract idea of information.[11] The third of the three points listed above will be addressed in the context of my sketch, to which I now turn.

Wittgenstein begins the *Philosophical Investigations* with a passage from Augustine's *Confessions* in which Augustine imagines how he learned language as a child: 'When they (my elders) named some object, and accordingly moved towards something, I saw this and I grasped that the thing was called by the sound they uttered when they meant to point it out. Their intention was shown by their bodily movements, as it were the natural language of all peoples ... Thus, as I heard words repeatedly used in their proper places in various sentences, I gradually learned to understand what objects they signified; and after I had trained my mouth to form these signs, I used them to express my own desires.' Immediately following, Wittgenstein comments: 'These words, it seems to me, give us a particular picture of the essence of human language. It is this: the individual words in language name objects – sentences are combinations of such names. In this picture of language we find the roots of the following idea: Every word has a meaning. This meaning is correlated with the word. It is the object for which the word stands' (§1). As Wittgenstein points out later on, Augustine's picture, that all words in language name objects, may arise from thinking only of words like 'table,' 'chair,' and 'bread,' or of people's names. If taken literally, the picture does not fit words like 'walk,' 'run,' 'if,' 'not,' and many others. Nonetheless, Augustine's picture has the form of another, closely related, and deeply seated idea: all words of language name objects of some kind, even if they may not be objects like stones, dogs, chairs, and tables. Why? Because every word has a *meaning*. Meanings are imagined as a special kind of object: one which can be correlated with *any* word. Thus the deeper idea underlying Augustine's picture retains its *form*: all words stand for *things*, after all. A word's *meaning* is the *thing* for which it stands. Meaning must be an abstract kind of thing, but since words have them, and since they are correlated with words, they must – somehow – *be there*. Finding out more about them awaits the exercise of our theoretical imaginations, but the picture of their existence, however shadowy, precedes and underlies explicit attempts to develop theories of meaning. G.P. Baker and P.M.S. Hacker point out that 'Augustine's conception of language is an *Urbild*. The family of philosophical accounts of meaning that grow out of it are full-blown "theories"' (1983, 4). Theories of meaning are fragile and vulnerable: they invite theoretical disputation. An *Urbild* works at a deeper level: it is harder to uproot.

Wittgenstein contrasts the Augustinian picture to a very simple, imagined use of language, or language game: 'Now think of the following use of language: I send someone shopping. I give him a slip marked "five red apples." He takes the slip to the shopkeeper, who opens the drawer marked "apples"; then he looks up the work "red" in a table and finds a colour sample opposite it; then he says the series of cardinal numbers – I assume that he knows them by heart – up to the word "five" and for each number he takes an apple of the same colour as the sample out of the drawer.' Immediately following this description of the language game, Wittgenstein presents an exchange with an imagined interlocutor: 'It is in this and similar ways that one operates with words. – "But how does he know where and how he is to look up the word 'red' and what he is to do with the word 'five'"? – Well, I assume that he *acts* as I have described. Explanations come to an end somewhere. – But what is the meaning of the word "five"? – No such thing was in question here, only how the word "five" is used' (§1).

Several important ideas are at work here. First, we encounter an idea central to Wittgenstein's later philosophy: that with descriptions of ways of operating with words, we have reached bedrock. The existence of our many and varied practices with words, practices whose interweavings spin the complex webs that make up our 'forms of life,' is a brute fact. Wittgenstein was fond of the line from Goethe's *Faust*, where Faust rewrites John 1:1 to read, not that in the beginning was the Word, but that in the beginning was the Deed (*Im Anfang war die Tat*): 'The origin and the primitive form of the language game is a reaction; only from this can more complicated forms develop. Language – I want to say – is a refinement, "in the beginning was the deed"' (1980, 31). When we accept that the shopkeeper's language game *is played*, we confront a second important idea of this simple example; that the abstract idea of meaning is not part of the picture. He emphasizes this point a few paragraphs later: 'Now what do the words of this language-game *signify*? – What is supposed to show what they signify, if not the kind of use they have? And we have already described that' (§10). The description of the practice with these words does not reveal a particular kind of object correlated with each word, and which answers the question: what is the meaning of that word? And yet – and this is the third important idea I want to draw from this passage – we are greatly tempted to say that simply acting as described cannot possibly be all there is to it. The word 'red' in the table looks nothing like a sample of *red*: how does the shopkeeper know that *red* is the meaning of 'red'? Is there not some intermediary – surely there *must* be one! – between the word 'red,' which consists after all in mere squiggles on paper, and the red of the apples fetched by the shopkeeper, something that the shopkeeper knows, and which explains how he can go from the word to the thing? And is that something not the *meaning* of the word?

One strand of Wittgenstein's multilayered response stresses the *occult* character of this abstract concept of meaning. When we take a statement like 'if the shopkeeper knows the meaning of the words "five red apples," then he knows how to use them,' not as a truism about the meaning of 'meaning' but rather as a philosophical insight awaiting further refinement, we are tempted to think that knowing what a word means consists in the presence of its *meaning* to the mind. Grasping the meaning is seen as *explaining* how it is that someone uses the word correctly. But Wittgenstein asks, if there is no problem with how the shopkeeper applies the word 'red' correctly when its *meaning* is before his *mind*, why should there be a problem of doing so when just the *word* 'red' is before his *eyes*? To insist that the problem simply does not arise for *meanings* is to impute magical or occult properties to them. Meanings are construed as entities that can reach right out to the world by a kind of mental projection or pointing. But the idea of a mental pointing that correlates a meaning present to consciousness directly with the world is a *superstition*: 'If the meaning of the sign (roughly, that which is of importance about the sign) is an image built up in our minds when we see or hear the sign, then first let us adopt the method ... of replacing this mental image by some outward object seen, e.g. a painted or modelled image. Then why should the written sign plus this painted image be alive if the written sign alone was dead? – In fact, as soon as you think of replacing the mental image by, say, a painted one, and as soon as the image thereby loses its occult character, it ceases to seem to impart any life to the sentence at all. (It was in fact just the occult character of the mental process which you needed for your purposes.)' (1969, 5).

Wittgenstein makes this point often. He says we are also tempted to impute occult properties to rules when we try to explain meaning in terms of rules governing the use of language. Imagined as anticipating what is in accordance with them, rules, like meanings, are magical. Regarding the +1 rule for extending the series, 1, 2, 3 ..., Wittgenstein remarks: 'The expression "The rule meant him to follow up 100 by 101" makes it appear that this rule, as it was meant, *foreshadowed* all the transitions which were to be made according to it. But the assumption of a shadow of a transition does not get us any further, because it does not bridge the gulf between it and the real transition. If the mere words of the rule could not anticipate a future transition, no more could any mental act accompanying these words' (142–3). Explanations, as Wittgenstein remarks in the example of the shopkeeper, come to an end somewhere. If we wish to know how one operates with words, we should look closely at the language games actually played. The important thing is what one does with the word, not something going on in the mind: 'There is a general disease of thinking which always looks for (and finds) what would be called a mental state from which all our acts spring as from a reservoir. Thus one says, "The fashion changes because the taste of people changes." The taste is the mental reservoir. But if a tailor today designs a cut of dress different from that which he designed a year ago, can't what is called his change of taste have consisted, partly or wholly, in doing just this?' (143). To operate with words is to engage in practices that precede reasoning and cerebration. The interpolation of a *meaning* between a word and its correct use is otiose. The effect of Wittgenstein's confrontation of the Augustinian picture with the shopkeeper's language game is to dissolve or deconstruct the abstract idea of meaning, not to offer a 'correct' *theory* of meaning. His investigation is intended to operate at a level where the abstract idea of meaning is exposed as a fiction.

The conception of meanings as occult intermediaries connecting word and world leads to the third of my three points about the value of Wittgenstein's investigation of the Augustinian picture. The language game of the shopkeeper helps us think about information because it involves the use of *documents*: in this case, a slip of paper, a label on a drawer, and a table correlating colour words with colour samples. In addition, it points to simple practices that intersect with uses of words, such as reciting the series of cardinal numbers. Many of Wittgenstein's language games feature diagrams, charts, tables, geometrical figures, and other kinds of documents. He shows us people engaged in practices with these things (see especially Part I of *The Brown Book*, 77–125).

His treatment of the Augustinian picture shifts our attention not only away from mentalistic pictures of meaning and towards practices with language, but also away from mentalistic pictures of information and towards practices with documents. If the abstract idea of meaning, which is meant to give life to dead, meaningless scribbles, is a fiction, then so too is the abstract idea of information, which is meant to give life to *documents*. The life of documents, like the life of words, is found in the uses made of them. Just as practices with words give words their life, practices with documents give documents their life. A document is no more informing because it embodies, or conveys, something called 'information' than a word is meaningful because it is correlated with something called a 'meaning.' Insofar as the Wittgensteinian perspective heralds a shift from *theories of meaning* to *descriptions of uses of language*, it also heralds a shift from *theories of information* to *descriptions of documentary practices*. One of the aims of this book is to show how rich and varied the practices with scientific documents can be, especially compared to the simplistic idea that there is no more to the informativeness of a document than what happens in the mind of someone who understands it.

The Plan of the Book

Representationalist views of science underwrite a parallel view of scientific documentation. In this book I suggest that there are formidable challenges to a view of science that privileges representation, theory construction, the coherence of conceptual fields, scientific rationality defined by scientific method, and an associated, abstract concept of information. There are, in other words, alternatives to what I call in this book *epistemological discourses of science*. An alternative now almost thirty years old, and which has generated a large and active literature, replaces the epistemological discourse of science with talk about material rather than conceptual practices. It does not confine itself to the intellectual products of science, but considers instead the matériel of the laboratory and the many labour processes involved in experimental science. This approach, whose intellectual precursor was Ludwik Fleck's *Genesis and Development of a Scientific Fact* (1979; 1st ed., 1935), a book which remained unnoticed until recuperated by Thomas Kuhn in the 1960s, received its impetus from the publication of David Bloor's *Knowledge and Social Imagery* (1976). Having evolved into an interpretation of the natural sciences in terms of the labour of stabilizing interrelationships between complex assemblages of heterogeneous elements, this approach emphasizes disunity rather than conceptual coherence; material things rather than abstract representations; practical and situated reasoning rather than an idealized scientific method; a wide variety of scientific activities at least as, if not more, important than theory production; and the significance of locality, contingency, situatedness, and historicity rather than a final and enduring stability of theoretical representations of the natural world. My approach unabashedly follows Hacking: 'The harm comes from a single-minded obsession with representation and thinking and theory, at the expense of intervention and action and experiment' (1983, 131).

I argue that in this alternative discourse, the ideas of information and its communication lose their centrality to analyses of the role of documentation in scientific activities. When the emphasis is on material practices and labour processes, the relevant questions about documentation shift from those interrogating its significance as a vehicle or channel for the communication of scientific information to those about the place of writing – and especially about the place of science's formal literature – among the variety of labour processes that make up scientific practices. Thus a reorientation of the view of science from epistemology to practice allows us to rethink the significance of this most curious of literary forms. The role of the scientific paper becomes much less paradoxical in a discourse of practice than in epistemological discourse.

But unravelling the paradox is not the book's main purpose. My intent is rather to use recent studies of the materiality and labour processes involved in scientific work as resources for reorienting studies of scientific information towards scientific documentation. If we can obtain a clear understanding of the many things scientists do, we can begin to ask fruitful questions about the place among them of the production and use of documents. Unravelling the paradox of the journal article is just one way to help us think more clearly about the role of writing in scientific practices. My aim, however, is broader: to propose a framework for understanding scientific literature in terms of documents, documentary circuits, and documentary practices.

Particular ontological assumptions underlie this framework. What exists are things and events: the things are documents, and the events are practices with them. Documentary practices are pursued in documentary circuits: the many and varied paths documents travel on the cultural terrain. Particular kinds of documentary practices are found in scientific cultures, and documents produced in what Fleck calls the esoteric centre of scientific activities also have a life in exoteric zones, just as the documents produced in the latter regions affect work in the former. To describe and understand them, one needs to investigate how documents are produced, the uses to which they are put, their zones of circulation, and the effects produced by practices with them.[12] Information, however, exists only as an effect of the ontologically primary elements: documents and documentary practices. It has, therefore, only a secondary or derived ontological status; it is an effect of the relative stability of documentary practices. Once practices stabilize, information can emerge. When it comes to analysing the information of science information systems, the issue becomes the one framed by Nunberg's question: 'Ultimately, then, the question we want to ask is phenomenological rather than lexicographical: not, What does "information" mean? but rather, How is the impression of "information" constituted out of certain practices ...?' (1996; 114–15).[13] In this book, information is deflated in order to reveal something more basic: the labour of establishing and maintaining documentary practices. Deflating information implies a rejection of the epistemological discourses of science that support the idea of scientific documents as mere vehicles for the communication of information.

A caveat must be entered about my use of 'epistemology' and its variants. In this book, I use the term more narrowly than its usual meaning. I refer solely to a conception of scientific knowledge as a theoretical representation of the world. Using it in this way has some historical justification, since the studies of scientific practices drawn upon here set themselves in opposition to the kinds of science studies pursued by philosophers, whose concerns are primarily epistemological in the traditional sense of looking to contexts of justification rather than contexts of discovery, focusing on justifying truth claims of scientific propositions, especially those of high theory. Be that as it may, the phrase 'epistemological discourses of science' is used here to refer to ways of speaking about science that privilege thought, rationality, conceptual fields, and production of theoretical representations of the natural world.

The book proceeds in the following steps. In chapter 1, three historical precursors set the stage for discussion of recent studies of science information systems. They are selected because, in spite of the importance each attaches to the social character of the scientific enterprise, they demonstrate the tenacity of epistemological assumptions by commitments to the *epistemic content* of documents as their primary and most significant feature when it comes to explaining their role in knowledge production. The first is Bacon's late sixteenth- and early seventeenth-century view of the social and political organization of scientific work. For him, knowledge production requires that the state organize and manage a complex institutional apparatus for the production and processing of scientific documentation. Yet his organization of scientific labour is nonetheless determined by the conceptual organization of scientific documents – what we would today call the classification or subject organization of scientific *information* – because his document organization derives from the order of knowledge itself. Epistemological principles thus configure Bacon's documentary, political, and social order of scientific labour. The second historical example is the work of Otlet, whose late nineteenth- and early twentieth-century documentalism exhibits a deep ambivalence between conflicting allegiances to documents or to *information* as his primary theoretical object. The former support an emphasis on the social order required to stabilize his 'universal book' as a source of knowledge, whereas the latter are embedded firmly in epistemological assumptions about scientific labour as production of the theoretical representation of a reality encompassing both the natural and social realms. The third example is the mid-twentieth-century work of Merton and his followers, whose pioneering sociological studies of a shared scientific ethos showed that science was much more than a cognitive activity. Yet the Mertonian norms governing that activity map onto epistemological assumptions about scientific rationality, lest the rewards distributed through the system of scientific documentation become detached from the discovery of scientific truth. Since the vehicle of reward is peer-reviewed publication, the formal literature becomes central to the scientific enterprise. But studies of the very limited use of science's formal literature suggest that Merton's ethos of science introduces an element that threatens to destabilize the marriage of science's cognitive and social orders.

These three brief historical sketches are followed in chapter 2 by an investigation of some scholarly studies of science information systems and information use among scientists. I argue that the significance to scientific activities of particular document forms, such as the journal article, has depended upon the idea of information as the epistemic content of documents, an idea congenial to epistemological discourses of scientific labour in which information acquires both theoretical coherence and the ontological properties of the substance sought as raw material of scientific theory.[14] Epistemic narratives of information are also traced in recent work on 'information seeking' and 'information behaviour.' An underlying theme is the degree to which such research exhibits realist or nominalist inclinations with respect to the appropriation of information as its leading trope. Epistemological interpretations of information are shown to expose the journal article to the conflicting and contradictory tensions of its uneasy, paradoxical status.

Chapter 3 introduces two analytical resources: (1) the distinction between an epistemological discourse of science and a discourse of scientific practices; (2) a model of activities and phenomena in which information and communication are central, which I call an information and communication model, or an model for short. Information and communication models supply the discursive resources for articulating science such that epistemological conceptions of information as a theoretical kind become central to it. But when science is articulated by an alternative set of discursive resources, which feature practices rather than theoretical representations, IC models are undermined. Information and communication lose their importance to analyses of science, and to analyses of the role of documents in it. The chapter introduces a central idea of the book: *deflating information*. Following Joseph Rouse's (1996) deflationary account of knowledge, I argue for a similar, deflationary account of information, thus clearing the way for a reorientation of the role of documents in science from the perspective of recent research on scientific practices.

Chapter 4 identifies some leading features of recent studies of scientific practices (SSP) that present special problems for IC models of science. Chief among them are the 'patchiness,' or 'motley' of the sciences, their disunity, and the heterogeneity of materials, processes, persons, groups, institutions, and techniques involved in experimental scientific labour. The selection of issues from the SSP literature is selective rather than exhaustive, as my primary aim is to identify those aspects most fruitful for rethinking information in terms of documentary practices. Thus the chapter considers differences in scientific discourse; the varied relationships between theory, observation, and experiment; the differences between information and laboratory resources; the contingency and temporality of scientific labour; the importance of documents in disciplinary agency, or what Pickering calls the 'mangle of practice'; and the porous boundaries between the 'internal' and 'external' phenomena of science. In each case, the point of the investigation is to trace implications for a documentalist approach to the study of science information systems.

Chapter 5 narrows the focus of SSP's view of science to the concern for formal writing and its place in scientific work. To this end, the chapter identifies some leading ideas on literature production from the SSP movement with particular relevance to the analysis of scientific documentation. The textual interpretations of science offered by Jerome R. Ravetz and by Latour and Woolgar are followed by Karin Knorr-Cetina's analysis of the relationship between the journal article and laboratory work and her interpretation of the article as a 'script of resource conversion.' Steven Shapin and Simon Schaffer's concept of the literary technology of science as an objectifying resource, Rouse's insistence on the narrative structure of scientific labour, and Fleck's interpretation of the different roles of the various strata of scientific literature are mined for analytical resources that, I argue, are more useful for understanding the role of scientific writing than the concept of the communication of information. One very general conclusion that emerges is that the labour of formal writing in science is best understood as the creation and maintenance of a set of objectifying resources for the articulation of local phenomena as universal features of nature. Such resources are critical to the historically contingent cultural form of Northern/Western science. This form is characterized by universalism rather than particularity, and by truth as adequacy of representation. In short, this cultural phenomenon is articulated just as epistemological discourses describe it. But, I argue, this cultural objective is an ongoing project rather than an accomplishment.

Chapter 6 considers in more detail documentary practices that reveal the historical contingency of modern science's universalizing project. It begins with some reflections on the contingency and history of universality, contrasting Aristotelian science to that produced in the early period of the Royal Society. But differences in forms of universality also appear in the same historical periods, as evidenced by the contrast between the cultures of basic science and engineering. Five sets of resources with particular relevance for documentation studies are identified: anti-representationalism, anti-essentialism, the process of cultural articulation, rejection of the idea of practices internal to science itself, and reflexivity.

Chapter 7 moves the discussion further along the path from science studies to documentation. Specific directions for research in scientific documentation and documentation generally are proposed: (1) translating talk about ideas, concepts, and information into talk about occasioned utterances and inscriptions; (2) interpreting inscriptions as actants, that is, as among the elements exhibiting agency and resistance in scientific labour; (3) expanding the range of scientific activities in which documents might play a role beyond legitimate, authorized science; (4) expanding the range of documentary circuits that figure in scientific labour; and (5) recuperating the role of history in documentation studies. The chapter concludes with some reflections on the role of science's organizational literature in stabilizing its formally documented inscriptions, and on how the modern science library helps construct the 'fantasia' of the objectivity and universality of scientific knowledge.

1

Epistemic Narratives of the Social Life of Science

How we think about science and how we think about scientific documentation have been inextricably linked at least since the development of the 'literary technology' of modern science in the middle of the seventeenth century (see Shapin and Schaffer 1985). It was not always so. A.C. Crombie's (1994; 1996) magisterial histories of the origins of our Western scientific traditions show that robust styles of scientific thinking long predate systems of scientific documentation. But certainly today, when we think about science we think at the same time of systems of scientific documentation, or – to use the contemporary preferred term – systems of scientific *information*. The reverse is also true. If we investigate descriptions of and proposals for science information systems, we encounter embedded assumptions about the nature of science. Our conceptions of scientific documentation (or information) intersect with our ways of thinking about science and what we take to be paradigmatic scientific activities.

This chapter explores connections between conceptions of science and its documentation in the thought of Bacon, Otlet, and Merton. These three are selected because each pays special attention to non- or extra-epistemic factors in knowledge production. All insist on specific social arrangements as necessary to scientific activity. Yet in each case, their thinking about systems of scientific documentation displays the strong epistemological commitments of a 'thought and theory' approach to the nature of science. The stories they tell about documents – Bacon's descriptions of the organization of documentary labour, especially in his Tables of Discovery and utopian model of the scientific state; Otlet's 'monographic principle' and celebration of the Universal Book; Merton's commitments to the importance of refereed publications to the social norms of a scientific ethos – are epistemic narratives. Underlying each of these stories is the idea that the feature of a document most important to knowledge production is its epistemic content. Whether it be called 'information,' epistemic content is an exemplar of the abstract, immaterial, conceptual substance that plays such an important role in representationalist, theory-dominated conceptions of scientific work. The epistemic content of a scientific document is what we grasp when we understand a sentence, diagram, graph, data set, computer-generated image, or any truth-telling inscription in any media form. It has the properties of wholeness, completeness, singularity, and presence. It is all there, all at once; each item of epistemic content is just what it is, distinguishable from and capable of standing in logical relations to any other. Since it is what we understand when the document makes a claim about the natural world, epistemic content has propositional form. It guides our imagination to a picture of scientific documentation as a conceptual system – a view most congenial to the epistemological view of science. The discussion below demonstrates the persistence of epistemic narratives, even in stories of the social life of science.

Cognitive Order: Bacon's Tables of Discovery

Bacon's model of properly conducted science emerges in the early seventeenth century from the fragments of his projected *Instauratio Magna*, the 'Great Instauration,' or renewal of the sciences. Bacon's thought is thoroughly materialist. His rejection of Platonism and Aristotelianism was grounded in an aversion to what he saw as his predecessors' rush to abstraction. Platonic forms, especially the Form of the Good from which all knowledge and existence flow, and the Aristotelian supreme Final Cause were both seen by Bacon as figments of a logical imagination transfixed by axioms from which conclusions about the natural world are deduced – a quest Bacon describes as the mind's natural tendency to fly 'from the senses and particulars to the most general axioms ... the truth of which it takes for settled and immovable' (1960, 43; Bk 1, §19). His new method of induction was meant to counter the tendency of the understanding to find rest in pure contemplation, but in order to do so, the understanding must first be disciplined by purging it of the false idols that lead it to 'just glance at experiment and particulars in passing' rather than dwelling 'duly and orderly among them' (43). 'The understanding,' he writes, 'must not therefore be supplied with wings, but rather hung with weights, to keep it from leaping and flying' (98).

Not only does the understanding's predilection for pure contemplation contaminate science with vague and insupportable generalities, logical fantasies, false phantoms, and superstition, it also subverts the true purpose of a properly conducted, materialist, experimental science. The aim of science is 'arts' and 'works,' terms Bacon uses to refer to the products of science applied to the improvement and ease of human life. Citing the *Cogitata et vista*, one of Bacon's early works, F.H. Anderson writes: 'The end and test of sound science is not contemplation but operation. Its pledge is the inventions which it provides – printing, gunpowder, the compass, engines, silk, and the like; "for truth is shown and proved by the evidence of works rather than by argument, or even sense"' (1948, 32). A chief fault of Aristotelian deductive science is its lack of any logic of discovery of new inventions: 'It cannot be that axioms established by argumentation should avail for the discovery of new works ... But axioms duly and orderly formed from particulars easily discover the way to new particulars, and thus render sciences active' (Bacon 1960, 44; Bk 1, §44). Bacon's materialism was therefore not merely philosophical; it subordinated science to practical ends: 'on account of the pernicious and inveterate habit of dwelling on abstractions it is safer to begin and raise the sciences from those foundations which have relation to practice, and to let the active part itself be as the seal which prints and determines the contemplative counterpart' (122–3; Bk 2, §4).

Given the practical dimensions of Bacon's materialism, his insistence on the importance of documents to scientific work comes as no surprise. Even though the genesis of the modern system of scientific literature is usually dated some forty years after Bacon's death in 1626, he placed writing at the centre of his vision of the social production of scientific knowledge, paying special attention to the production, collection, organization, processing, and systematic use of documents. The best-known exemplar of his concern for documentation is his classification of the sciences, as first developed in the *Advancement of Learning* of 1605. This classification – 'The Divisions of the Sciences' – is described in the 'Plan of the Great Instauration' of the *Novum Organum* of 1620 as the first of the Instauration's six parts. Bacon describes it in documentary terms: 'a *summary* or *general description* of the knowledge which the human race at present possesses' (17; emphasis added). It is therefore more than an abstract framework for thinking about how the various sciences are related to one another: it is a *document* classification. Its most fundamental classes reflect 'the absolute chasm which exists between the truths given in revelation through the Word of God and axioms discovered by the powers of man and, secondly, through distinctions among the human faculties' (Anderson 1948, 148–9).[1] The three main classes, which correspond to the three main human faculties – memory, imagination, and reason – are history, poetry, and philosophy. History is divided into natural and civil. The natural history class documents the 'phenomena of the universe' in a systematic organization of records 'not only of nature free and at large (when she is left to her own course and does her work in her own way) – such as that of the heavenly bodies, meteors, earth and sea, minerals, plants, animals – but much more of nature under constraint and vexed; that is to say, when by art and the hand of man she is forced out of her natural state, squeezed and moulded' (Bacon 1960, 25). The causes of natural phenomena belong to physics ('variable and respective causes') and metaphysics ('fixed and constant causes'), both classed under natural philosophy. In characteristic Baconian fashion, these 'divisions of the sciences' are not intended merely for contemplation of the interrelations of the various sciences, but as an instrument for organizing documented knowledge.

Bacon's classification involves much more than simply slotting scientific documents into the categories of the divisions of the sciences. Unlike modern library classifications, which classify documents that just happen, one way or another, to get published, Bacon's scheme sits atop a complex and highly regimented social system of document production. That system, implied throughout the fragments of the Instauration, is most vividly conveyed in *New Atlantis*, Bacon's utopian fable of state-organized science, where he imagines scientific document production as socially disciplined by the goal of scientific knowledge production.[2] Things 'already invented and known' are collected principally from books and other written records. This document collection activity is divided among several different ranks of scientific worker: Merchants of Light (those who sail to distant lands to collect reports of experiments), Depredators (those collecting local experimental reports), and Mystery-Men (those collecting reports of the mechanical arts). These documents of natural history are then processed by further ranks of scientific worker. The Compilers extract 'titles and tables' from the collected records in order to display observations in perspicuous fashion, thus allowing 'axioms' to be more easily inferred from them.

Reflecting upon the tables produced by the Compilers and after consultation with other scientific colleagues, the Lamps suggest new experiments to advance knowledge by building upon previous and current work. Such experiments are then performed by the Inoculators, who submit their reports to the Interpreters of Nature. The task of the Interpreters is to generate higher-order axioms to direct further observations. Finally, the Dowry-Men, or Benefactors, apply the knowledge of nature to the production of useful inventions. In the divisions of the sciences, the reports of the Lamps, Inoculators, Interpreters of Nature, and Benefactors belong to philosophy. Philosophy is the third and highest class in the Division of the Sciences; it includes the natural knowledge of God, man, and nature. The work of the Benefactors belongs to operative philosophy, and that of the Lamps, Inoculators, Interpreters of Nature to speculative philosophy, each a branch of natural philosophy. Thus the division of the sciences functions as a blueprint of the machinery of socially organized knowledge production.

Bacon's view of scientific knowledge production is in many ways more subtle and complex than overly intellectualized views which equate scientific activity with the production of conceptual fields. His insistence on the social organization of scientific labour, on strict rules for writing scientific documents, and on applying science to invention in the production of machines, instruments, and technological apparatus resists the reduction of scientific activity to cognitive operations – a view derided by Andrew Pickering as one of 'disembodied intellects making knowledge in a field of facts and observations' (1995, 6). Moreover, political concerns figure prominently in Bacon's system. Although the manufacture of knowledge emphasizes the proper production and organization of written records, such records are not to be made public. Fearful of the social and political dangers of widespread natural knowledge to the security of the imperial state, Bacon granted the fabled brethren of his utopia the exclusive right to control publication of their reports (Martin 1992, 138–9).

Yet in spite of Bacon's strong materialism, his acute awareness of the social organization of scientific labour, and his close attention to the need for a systematic and disciplined production of scientific documents, his documentary system is nonetheless driven by epistemological imperatives deriving from the centrality to it of a document's epistemic content. In his system, what is important about documents is what they convey. Science can be built from a documentary system precisely because documents transmit the epistemic content needed to construct a model of the world in human understanding. Bacon says: 'I am building in the human understanding a true model of the world, such as it is in fact, not such as a man's own reason would have it to be; a thing which cannot be done without a very diligent dissection and anatomy of the world' (113). Such a model of the world is possible because the world is a unified whole: 'Men's labour therefore should be turned to the investigation and observation of the resemblances and analogies of things, as well as in wholes as in parts. For these it is that detect the unity of nature, and lay a foundation for the constitution of the sciences' (175–6). Perez Zagorin points out that as a 'materialist and a realist in the modern, philosophical meaning of the term, Bacon had no doubt that the world and nature existed independently of the mind. It followed that thought and language must give a true representation of the world if they are to qualify as knowledge' (1998, 79). Bacon's materialism recognized that thought and language need documents. But their whole purpose is to convey in carefully constructed language what thought needs to represent the world.

The first step in a system of scientific document production is patient and scrupulous recording of the phenomena of nature: 'For first of all we must prepare a natural and experimental history, sufficient and good; and this is the foundation of all, for we are not to imagine or suppose, but to discover, what nature does or may be made to do' (Bacon 1960, 130). Anderson remarks that this natural history is a 'delineation of the sort of experimental history which is suitable for the building of a philosophy.' Because 'it provides the materials on which the understanding is to operate,' this part of the classification is so necessary 'to the Instauration that, if it cannot be provided, the scheme cannot become operative and the whole project for the reform of knowledge may as well be given up' (259, 260). But natural history alone cannot supply what the intellect needs to model the world, because its phenomena are so many and diverse. Its records must be processed by literary techniques of condensation and compression, such that the understanding may be guided in its careful ascent from particulars to axioms. The kind of documents that aid the intellect in this way are Tables of Discovery: 'since there is so great a number and army of particulars, and that army so scattered and dispersed as to distract and confound the understanding, little is to be hoped for from the skirmishings and slight attacks and desultory movements of the intellect, unless all the particulars which pertain to the subject of inquiry shall, by means of Tables of Discovery, apt, well arranged, and, as it were, animate, be drawn up and marshaled; and the mind be set to work upon the helps duly prepared and digested which these tables supply' (Bacon 1960, 97). Bacon's metaphor is telling: the tables are 'animate'; they are, as it were, *alive*, because if properly prepared and arranged, they lead the human understanding to conclusions about the natures of things as if guided by a living being.

Bacon uses the example of heat to explain how the tables operate. The world is not a structure of particular things, which are infinite in number, but of the 'forms' inherent in them, which are finite in number, and 'which produce by their conjunctions natural things in infinite variety' (Anderson 1948, 217). Knowledge of the forms of 'natures,' such as heat, or redness, is gained by juxtaposition of several varieties of natural phenomena, or 'instances,' carefully chosen to lead the intellect to the law of the form in question. One begins with a Table of Essence and Presence, which lists a number of instances that agree in the respect of exhibiting heat, such as the rays of the sun, fiery meteors, burning thunderbolts, and 'green and moist vegetables, confined and bruised together, as roses packed in baskets; insomuch that hay, if damp, when stacked, often catches fire' (Bacon 1960, 131). The second step is a Table of Deviation, or of Absence in Proximity, which lists cases in the first table where the given 'nature' is absent, such as the observation that the 'rays of the sun in what is called the middle region of the air do not give heat ... And this appears from the fact that on the tops of mountains, unless they are very high, there is perpetual snow' (133). The third step is a Table of Degrees of Comparison in Heat, 'in which the nature under inquiry is found in different degrees, more or less' (142), such as: 'Let further inquiry be made into the different degrees of heat in different animals, as in fishes, quadrupeds, serpents, birds; and also according to their species, as in the lion, the kite, the man; for in common opinion fish are the least hot internally, and birds the hottest, especially doves, hawks, and sparrows' (144). From these three tables, which Bacon calls the Presentation of Instances to the Understanding, 'induction itself must be set at work': 'the problem is, upon a review of the instances, all and each, to find such a nature as is always present or absent with the given nature, and always increases and decreases with it; and which is, as I have said, a particular case of a more general nature' (151). Zagorin describes this as a process of 'eliminative induction': 'a dissection or separation by eliminating the several natures not found in some instances in which the given nature, heat, is present; or that are found in some instances in which the given nature is absent; or are found to increase or decrease in some instance when the given nature acts in the opposite manner' (1998, 98, 99). Conclusions are reached on the basis of the contents of documents; it is as if, once records of phenomena of nature are produced and properly condensed into suitably arranged tables, reflection on documents themselves performs the essential work of science.[3] Led by 'animate' documents, the understanding reaches conclusions about the natures of things by induction, the proper logic of scientific discovery. The aim of Baconian science – a model of the world in the human understanding – is a product of *thought* rigorously disciplined by documents. Properly produced and arranged, documents are the material vehicles of the abstract epistemic content thought needs to model the world. Bacon's story of the role of the tables in the production of scientific knowledge is thus an epistemic narrative of scientific documentation.

The work Bacon left to posterity was but a fraction of his Great Instauration. Many questions about different aspects of his project remain unanswered. A question of interest to documentation is, to what extent is Bacon's grand documentary scheme, as envisioned in his classification of the sciences, also an epistemic narrative? Put in another way, to what extent is Bacon's classification of the sciences a reflection of the model of the world in the human understanding? Several points bear on this issue.

First, Bacon's classification is an organization of scientific writing. The connection between disciplined scientific writing and its epistemic content is established in his treatment of the Tables of Discovery. Moreover, Bacon is aware of the importance of writing to science as a whole. In the course of answering an objection that he is 'but doing what has been done before,' he replies: 'For the ancients, too, it will be said, provided at the outset of their speculations a great store and abundance of examples and particulars, digested the same into notebooks under heads and titles, from them completed their systems and arts, and afterward, when they understood the matter, published them to the world, adding a few examples here and there for proof and illustration; but thought it superfluous and inconvenient to publish their notes and minutes and digests of particulars, and therefore did as builders do: after the house was built they removed the scaffolding and ladders out of sight.' Their failure to publish notes and digests, and their eagerness to remove the documentary scaffolding, were part and parcel of their greatest mistake: 'they flew at once to the most general conclusions, or first principles of science' (114; Bk 1, §125). The implication is that since his science proceeds methodically from carefully documented particulars, writing is essential to science as a whole. Elsewhere, Bacon argues for the importance of writing to natural history: 'But even after such a store of natural history and experience as is required for the work of the understanding, or of philosophy, shall be ready at hand, still the understanding is by no means competent to deal with it offhand and by memory alone ... And yet hitherto more has been done in matter of invention by thinking than by writing; and experience has not yet learned her letters. Now no course of invention can be satisfactory unless it be carried on in writing. But when this is brought into use, and experience has been taught to read and write, better things may be hoped' (96–7; Bk 1, §101). The importance of scientific writing is found in its epistemic content; writing is necessary for natural history generally; the classification of the sciences is an organization of scientific writing – these are reasons to interpret Bacon's classification of the sciences as epistemically driven.

Second, we have seen that Bacon emphasizes the need for a systematic grouping of the Tables of Discovery, referring to them as effective only when 'apt, well arranged.' He uses similar language in describing natural history as a whole: 'since my method of interpretation, after the history has been prepared and duly arranged, regards not the working and discourse of the mind only (as the common logic does) but the nature of things also, I supply the mind such rules and guidance that it may in every case apply itself aptly to the nature of things' (116; Bk 1, §115). Anderson points out the connection between induction and an orderly arrangement of the particulars of natural history in general: 'Induction rests on the foundation of *sorted* natural history' (219; emphasis added). This arrangement of the records of particulars would form part of the natural history class in the overall classification. It seems most likely that Bacon imagined this arrangement not as a special case of epistemic order in just one part of the classification, but as exemplifying the aid to intellectual work that the classification as a whole can perform.

Third, the need for 'arrangement' is not specific to the Tables of Discovery or to the natural history class as a whole, but essential to the method they facilitate. Induction is the way to any true knowledge: 'It may be asked ... whether I speak of natural philosophy only, or whether I mean that the other sciences, logic, ethics, and politics, should be carried on by this method. Now I certainly mean what I have said to be understood of them all; and as the common logic, which governs by the syllogism, extends not only to natural but to all sciences, so does mine also, which proceeds by induction, embrace everything' (115–16; Bk 1, §127). Since proper arrangement of documents leads the mind to inductive conclusions in any branch of knowledge, we see that Bacon's classification of all knowledge, whether of Nature, God, or Man, is driven by the same epistemic concerns as those at work in the Tables of Discovery.

Finally, Bacon clearly expresses the scope of his epistemological principles by remarking that in his Division of the Sciences, 'I take into account not only things already invented and known, but likewise things omitted which ought to be there' (18). He can include things 'which ought to be there' because, unlike the library classifications of the nineteenth and twentieth centuries, his classification reflects knowledge organization rather than just the practical business of document access.[4] It does not derive its warrant – the justification of the categories used in the classification system – from extant documents (i.e., it is not based upon literary warrant), but from an arrangement designed to guide the intellect according to the logic of true scientific reasoning to conclusions about the laws of the forms governing the 'natures' of things. This arrangement also drives the production of new documents, because once the understanding is led by the proper organization of knowledge from experience to axioms it derives from them new experiments, from which new experimental records are generated, thereby closing epistemic gaps in the classification.

The different aspects of Bacon's thinking about documents and their arrangement canvassed here, when combined, form a tentative yet reasonably coherent picture of what today we would call a science information system. It is a robustly social system. Bacon saw the scientific enterprise as a collaborative, socially organized activity. He recognized that science does not develop by observation and experiment alone, but requires a literature: science is as much about writing as it is about thinking. In the *New Atlantis*, Bacon envisioned a social order organized by scientific knowledge production. The institutional organization of documentary labour was central to it.

An epistemic order underlies and supports this social order, and to the extent that it does, Bacon's story of the social system of knowledge production is an epistemic narrative. If we superimpose his belief in the unity of nature, in the possibility of modelling the world in human understanding by applying the inductive method of reasoning, and in the essential role of duly prepared and properly arranged documents in leading the intellect from particulars to axioms, there emerges a picture of a documentary system whose organization reflects the order and coherence of the world and our knowledge of it. Disciplined production and use of documents are so important in the Baconian scheme because their primary function is to convey the epistemic content needed to model the world. And, since the highest purpose of the social order is to guide action by knowledge, the organizing principle of the institutional structures of the ideal state, as described in the *New Atlantis*, is also derived from the organization of knowledge. Documents, knowledge, the world, and the social order line up one behind the other, each reflecting each.

What would the structure of scientific documentation look like, given Bacon's view of the scientific enterprise? Reports of observations, either of nature 'free and at large' or 'under constraint and vexed,' constitute the basic level of the scientific literature. The second level consists in the records generated by processing those of the previous stratum in order to aid the intellect in reaching conclusions about 'natures,' such as heat: exemplars are the Tables of Discovery. The third level are documents recording conclusions, or 'axioms,' that support practical operations, inventions, and further experiments and investigations. The highest level record the most general axioms. Such a documentary structure is hierarchical, culminating in a few documents containing high-level generalizations. In this model, documentalists – those who oversee the production and arrangement of documents – are the co-workers of experimental scientists, and are therefore central to the production of knowledge. Bacon has provided us with a very robust epistemic narrative of a science information system.

Otlet, Science, and the Universal Book

Recent years have seen a steep rise of interest in the work of Paul Otlet.[5] Otlet (1868–1944) was a Belgian lawyer, bibliographer, and utopian internationalist. In 1895 he co-founded, with Henri La Fontaine, the International Institute of Bibliography, later known as the International Federation for Information and Documentation (it was dissolved as of 1 January 2001). His internationalism involved active participation in movements that led to the founding of the League of Nations and Unesco. Like Bacon, Otlet recognized the social character of information. Boyd Rayward, his biographer and foremost interpreter and scholar, writes, 'for Otlet the document is at the centre of a complex process of communication, of the cumulation and transmission of knowledge, of the creation and evolution of institutions' (1991, 137). He goes on to suggest that Otlet's *Traité de Documentation* (1934) 'is perhaps one of the first comprehensive introductions to the study of information as an important social phenomenon' (138). As Ron Day puts it, 'Perhaps there has never been an information theorist who took so seriously the mapping of information upon the social' (1997, 43). Rayward's collection of selected papers (Otlet 1990) reveals Otlet as the kind of writer who eschewed systematic theorizing in favour of promoting specific projects and organizations in aid of *documentation*, the name he gave to a new kind of professional, bibliographical practice and its associated institutions. Among the many strands of thought about science, documents, classification, and bibliography found in these papers, one in particular reveals a striking continuity with Bacon's ideas about science information systems.[6] In this aspect of his thought, Otlet, like Bacon, subordinated the social arrangements in which information is implicated to epistemological principles. As Rayward notes, although Otlet's theory 'considers documentation as a social phenomenon,' it nonetheless 'attempts to set it as firmly as possible in epistemology' (1967, 263). The centrality of epistemology is evident in the way he describes, in his early work, his notion of the Universal Book. As we will see, the Universal Book relies upon a highly reified conception of information as a unique, unified, and uniform kind of thing. Otlet's story of the compilation of the Universal Book, because it relies on the idea of the epistemic content of documents, is his version of an epistemic narrative of documentation.

Otlet contributed to the many efforts made in the 1890s by scientific associations to improve the bibliographic control and organization of scientific documents. In 1895, he and La Fontaine organized the International Conference of Bibliography, just a year before the Royal Society of London had planned to convene a conference on issues arising from its *Catalogue of Scientific Papers*. For the most part, as Rayward notes (1975), Otlet's and La Fontaine's efforts were resisted by bibliographers but encouraged by scientists, an unsurprising state of affairs given the enthusiastic scientism so evident in Otlet's early work. In 'Something about Bibliography,' written in the early 1890s, Otlet writes in a thoroughly Baconian spirit: 'The results of the natural sciences are grounded in millions of carefully observed, analysed, and catalogued facts. These facts have subsequently been integrated into sequences and the combination of these sequences has naturally led to the enunciation of laws,

partial at first, general later, from which the most powerful and indestructible synthesis that has ever been made now seems possible' (1990,
11). For Otlet, therefore, scientific labour was based upon, as it was for
Bacon, preliminary work of careful and meticulous gathering and
arranging of phenomena, observations, and experimental results. But
not just gathering and arranging: again like Bacon, these 'facts' must
also be carefully documented, as the reference to *catalogued* facts indicates. In the natural sciences, 'each new contribution ... seems to be
recorded immediately and to become for everyone the point of departure for future research' (12). Each fact is classed with like facts, and sets
of facts (or 'sequences') thus formed are classed with like sets at higher
levels of generality. Scientific rationality is so powerful, in Otlet's view,
that laws emerge *naturally* from this organizational labour – even without anything like Bacon's Tables of Discovery. These laws are hierarchi-
cally arranged, thus reflecting the ideal of unified science – 'the most
powerful and indestructible synthesis that has ever been made.' Scientists 'at any moment ... are always aware of the present status of their science'; they work on 'the same edifice the broad outline of whose design
has been impressed upon each one of them' (12). Otlet's ideas of careful
documentation of facts and their arrangement in the service of a scientific rationality that aims at a single, unified model of the world reprise
Bacon's epistemic narrative of scientific knowledge production. A striking dissimilarity between the two is evident in their radically divergent
conceptions of the adequacy of human understanding. Bacon's struggle
against the Aristotelian drive to abstraction emphasized purging the
understanding of false idols. For Bacon, the road to scientific knowledge
was strewn with obstacles. For Otlet, although the documentary work
of knowledge production was formidable, the generation of laws was
almost automatic once the proper documentary system was installed.
Unlike Bacon, he exhibited an energetic faith in the capacity of human
understanding to easily model the world.

Otlet recognizes that what he calls 'the great monument of the natural sciences' is only an ideal. Nonetheless, natural science provides the
only ideal, since all knowledge takes the form of natural-scientific
knowledge: 'Knowledge merely comprises all observed facts and all of
the likely hypotheses that have been formulated to explain these facts
and reduce them to laws' (17). If Otlet's vision of science seems positivistic, it was meant to be. He is quite explicit on this point: 'Natural
scientists are not content simply to declare themselves positivist as a
turn of phrase, as most of our popularisers do, and then to act as if pos-

itivistic methods did not exist *and ought not to be applied everywhere and always'* (11; emphasis added).[7] It comes as no surprise that Otlet could quite readily champion positivistic methods for knowledge production in the social sciences of his day, because the positivism available to him in the 1890s was not logical positivism, which came later and was concerned mainly with the natural sciences, but the positivism of Auguste Comte, who was concerned to establish 'positive' knowledge of social phenomena. Otlet's deep commitment to internationalism and world government, together with his robust scientism, led him to the inescapable conclusion that world peace and a just, global society depend upon rational thought in both the natural and social realms – and rational thought is simply scientific thought universally applied. The scope of the positivistic elements of Otlet's thought supported his conviction that science was the solution to a social crisis and a blueprint for a harmonious international society. Social facts, therefore, must be as rigorously gathered and as faithfully documented as natural facts. *Documentation* – the rigorous and perspicuous presentation of facts as evidence for scientific claims – is therefore essential to the success and progress of any and all true knowledge.[8]

Having established that all kinds of facts must be carefully gathered, observed, analysed, and catalogued, Otlet realized that how facts are catalogued – how they are *documented* – determines whether they are presented with the perspicuity required for the 'natural' evolution of scientific laws, or whether they are obscured, leading to ignorance and chaos. The documentary rigour of the natural sciences is exemplary, because their 'millions of carefully observed, analysed, and catalogued facts' have allowed laws to emerge 'naturally.' But things are quite different in the social sciences: the 'debasement of all kinds of publication resulting from the modern cultivation of the so-called moral, social and political sciences is alarming to those who are concerned about quality rather than quantity' (11). Social scientists are simply not positivistic enough, in comparison to their colleagues in the natural sciences. The social sciences therefore present far greater problems than the natural sciences in extracting the constitutive units of knowledge – the facts – from the 'debased' forms in which they are embedded but hidden from view due to the failure of the social sciences to practise the methodological rigour of the natural sciences.

The greatest challenge faced by documentation as a profession and necessity for science stems from a particular documentary form: the book. The problem with the book, especially in the 'debased' social sciences, where it is more highly favoured than in the natural sciences, is that it is all too often full of 'fine language or repetition or padding.' This dross obscures what is of genuine value in the book: *the facts*. 'The external make-up of a book, its format and the personality of its author are unimportant provided that its substance, its sources of information and its conclusions are preserved and can be made an integral part of the organisation of knowledge, an impersonal work created by the efforts of all.' But there is a remedy, just as there is a technique to pan gold from dirt. The remedy is *documentation*. A new kind of information worker – a *documentalist* – is envisioned by Otlet: one who delves inside the book to hunt down and extract 'whatever is new and adds to knowledge.' Otlet describes the documentalist's work: 'The ideal, from this point of view, would be to strip each article or each chapter in a book of whatever is a matter of fine language or repetition or padding and to collect separately on cards whatever is new and adds to knowledge.' This work is different from that of the librarian (the documentalist's predecessor), who accepts the form of the book as it is and tries to capture its overall content in a subject heading applicable to the work as a whole.[9] Doubtlessly recognizing the magnitude of the documentary task lying ahead, Otlet allows that 'such winnowing to conserve the best grain is clearly only an ideal' (17).

The principle by which the essential units of knowledge may be extracted from books, and therefore the principle governing documentation and the work of documentalists, is called by Otlet the *monographic principle*. The assumptions of the monographic principle are: (1) there is such a thing as information; (2) it is recognizable in itself, and therefore extractable; (3) its criterion of identity is epistemic; (4) information is most valuable when its boundaries are sharply visible, so that it achieves presence as a unit of knowledge; and (5) information gains value through the work of documentalists, who are trained to recognize units of knowledge and extract them, even from documentary forms such as the book, which threaten to bury them under the subjective and personal padding introduced by their authors. In his scientist posture, Otlet presented individual subjectivity as a source of contamination. What is valuable about a piece of writing derives from universal features of rationality as such, not from individual differences between human subjects. As Rayward elegantly puts it, documents are 'processed in such a way that those wanting information from them were released from the tyranny of sequence and structure imposed by the authorial act of creation' (1994a, 170).

The documentalist, like the scientist, is a hunter and gatherer of facts. The scientist hunts facts in both nature and documents; the documentalist hunts only in documents. But in either case, simple fact-gathering is not enough. Knowledge can be generated from facts only if the facts are properly classified. At first, Otlet thought that the facts recorded on cards could easily be organized: 'These cards, minutely subdivided, each one annotated as to the genus and species of the information they contain, because they are separate could then be accurately placed in a general alphabetical catalogue updated each day' (17). He quickly saw that an alphabetical organization would be inimical to the catalogue's effective use in knowledge production. The challenge is to classify documents in such a way as to enable the 'natural' organization of facts into 'sequences,' so that laws are generated 'naturally' from them: 'Once facts are observed and recorded in publications, thus becoming an integral part of knowledge, we would simply like to find out if with a special classification it is possible eventually and naturally to group them into scientific laws' (12). Since the organization of facts involves progressive generalization, the most fundamental categorization is by genus and species. Therefore to take facts out of the context 'imposed by the authorial act of creation' and to rearrange them according to this most basic category of scientific thought enables knowledge production: 'For written works a rearrangement of their contents not along the lines of the special plan of a particular book, but according to the genus and species appropriate to each element does not make for any loss of substance' (17). An effective organization of records of facts must be based upon the categories of scientific reason itself. A well-ordered set of documents becomes an organ for knowledge production; Otlet therefore characterizes his project as 'the creation of an artificial brain by means of cards containing factual information' (17).

Otlet was inspired by the Dewey Decimal Classification (DDC) as the basis for a truly scientific organization of elementary facts. He was attracted not to the verbal categories of the system, but to what he regarded as its 'scientific' notation. Theoretically, Dewey's system of ten basic categories generates, by means of its nested sets of ten subcategories capable of being repeated *ad infinitum*, an infinitely expandable system of subject division and subdivision. Otlet writes that it 'provides for human knowledge a nomenclature which is stable and universal and can be expressed in an international language – that of numbers' (26). He goes on to describe it in a manner very foreign to

Melvil Dewey's practical way of thinking: 'the *Decimal Classification* actually constitutes an international scientific language, a complete system for representing science which one day perhaps may bring help to intellectual workers analogous to that which they received from Latin in the Middle Ages and during the modern period' (34). Otlet sees in Dewey's notational elements a rigorous language for designating concepts: 'they express abstractions, pure scientific categories ... they translate ideas absolutely common to the entire scientific world and express them in universally understood signs – numbers' (34). He therefore adapted and greatly enhanced the expressiveness of the DDC to create, with La Fontaine, the Universal Decimal Classification (UDC). Ideally, its purpose was to impose on facts the organization they need to become the raw materials processed by the 'artificial brain' to generate scientific knowledge. The monographic principle and the UDC became Otlet's software for knowledge production.

When facts are extracted from books, and from each author's personal preferences of order and arrangement, their Universal Decimal class notation places them in the 'sequences' required for knowledge production. The facts therefore migrate from many scattered individual and personal books to one great, collective, and impersonal book: the Universal Book. It consists of all the facts, recorded on cards, together with their classification according to the UDC. He describes it as follows: 'Information, from which has been removed all dross and foreign elements, will be set out in a quite analytical way. It will be recorded on separate leaves or cards rather than being confined in volumes ... By gathering these leaves together, and classifying and organising them according to the headings of a reliable, precise, and detailed classification, we will create the 'Universal Book' of knowledge, a book which will never be completed but will grow unceasingly' (84).

Otlet's story of the Universal Book is an epistemic narrative writ large, and perhaps the most radical reification of information before our contemporary musings about digital information as pure thought. The monographic principle, according to which a book's 'facts' are extracted without loss – indeed, with great gain – displays in starkest relief the idea of *information* as the epistemic content of documents. The criterion of identity for Otlet's 'facts' is *epistemic*: if it contributes to *knowledge*, then it is extracted and recorded on a card. The work of the documentalist might be imagined as panning for epistemic nuggets in books. The production of the Universal Book – a massive collection of cards, each recording a nugget of information – by expertly trained cadres of documentalists is an image of the world's books decon-structed, only to be reconstructed in a Book whose purpose and essence are purely, and rigorously, epistemic. This infinite, never-ending Book to end all books encompasses all knowledge, and disciplines writing itself in its own image. Otlet's description of it is rhapsodic:

> paralleling the innumerable books published on the subject-matter of each discipline, will be drawn up the 'Universal Book' of that discipline. This Book, the 'Biblion,' the Source, the permanent Encyclopedia, the Summa, will replace chaos with a cosmos. It will constitute a systematic, complete and current registration of all the facts relating to a particular branch of knowledge. It will be formed by linking together materials and elements scattered in all relevant publications. It will comprise inventories of facts, catalogues of ideas and the nomenclature of systems and of theories. It will condense various scientific data into tables, diagrams, maps, schemas. It will illustrate them by drawings, engravings, facsimiles, and documentary photographs. It will be like a great cadastral of learning, in which all elements in knowledge will be reported and recorded day by day ... The old form of books will no longer be maintained; they must give way before the abundance and variety of matter. (83–4)

In Otlet's later work we encounter the Universal Book in a form that makes all the world's knowledge available to everyone by combining printed documents with photography, miniaturization through micro-forms, the use of film, radio, television and even X-rays. He writes that a 'radical assumption would consider that all knowledge, all information could be so condensed that it could be contained in a limited number of works placed on a desk, therefore within hand's reach, and indexed in such a way as to ensure maximum consultability. In this case the world described in the entirety of books would really be within everyone's grasp' (1). As Rayward (1994b) has noted, it is the technological fantasy of a World Brain, in which all knowledge is gathered in one place. Otlet supplements his World Brain vision with another, one utterly familiar to us today, of thinking machines absorbed, cyborg-like, into the human body, an extended organ of thought, in a gesture that Rayward sees as a precursor of the imaginary space of today's fantasia of disembodied, digital information systems. In Otlet's utopian dream, the 'Universal Book created from all books would become very approximately an annex to the brain, a substratum even of memory, an external mechanism and instrument of the mind but so close to it, so apt to its use that it would truly be a sort of appended organ, an exodermic appendage' (1990, 1). His epistemic narrative culminates in a vision of the Universal Book as partaking of the divine, a phenomenon of the fall of man only insofar as it manifests to mortals through material and technological means an image of the cosmos always present to divine consciousness:

> Man would no longer need documentation if he were to become an omniscient being like God himself. A less ultimate degree would create an instrumentation acting across distance which would combine at the same time radio, x-rays, cinema and microscopic photography. All the things of the universe and all those of man would be registered from afar as they were produced. Thus the moving image of the world would be established – its memory, its true duplicate. From afar anyone would be able to read the passage, expanded or limited to the desired subject, that could be projected on his individual screen. Thus, in his armchair, anyone would be able to contemplate the whole of creation or particular parts of it. (1)

The manifestation of a comparable apotheosis of information had to await the cyber-enthusiasms of the very late twentieth century.[10]

Cognitive Contamination: Merton's Ethos of Science

Both Bacon's and Otlet's stories of scientific knowledge production emphasize its embeddedness in documentary institutions. Yet both are epistemic narratives. In each case, social order is subordinated to epistemological and cognitive order, because the work of science is viewed as the construction of a systematic and unified conceptual field. Documents are central to this work because they convey raw material for scientific labour in the form of epistemic content. But for Merton – the founder of the sociology of science – epistemic imperatives alone are insufficient to organize scientific knowledge production. He sees science as a self-regulating social order whose cohesiveness, stability, and systematic advance depends upon shared values manifested by adherence to specific norms – an 'ethos of science ... that affectively toned complex of values and norms which is held to be binding on the man of science' (1973d, 268–9). Since Merton's norms are treated in detail in an extensive literature, it suffices here to simply list his original four norms: organized scepticism (scientists are expected to evaluate claims to new knowledge critically and objectively); disinterestedness (their findings are not expected to be used in a self-interested fashion; they are expected to maintain an attitude of emotional neutrality toward their work); universalism (scientific merit should be evaluated independently from the personal or social qualities of individual scientists); and communalism (since scientists do not own their findings, secrecy is forbidden, and open communication is prescribed).[11]

Since Merton's social norms build moral imperatives into the heart of scientific activity, knowledge production comes to depend upon more than cognitive and technical standards. The scientist must not only follow rigorous methodological precepts, such as those Bacon took great pains to elaborate, but also work at 'fashioning his scientific conscience' (269). But when the social system of science is made to depend upon faith as well as reason, then Merton is faced with a problem Bacon did not have: to ensure that keeping the faith and acting rationally do not conflict. The ethos of science must map moral standards onto the standards of scientific rationality; the social system of conscience building cannot be incompatible with or subvert the rigorous epistemic and cognitive tests of genuine scientific work.

Merton's solution to the problem of integrating the moral with the cognitive and epistemic imperatives of science rests on his claim that the norms preserve the integrity of science because they flow from scientific method:

> The institutional goal of science is the extension of certified knowledge. The technical methods employed toward this end provide the relevant definition of knowledge: empirically confirmed and logically consistent statements of regularities (which are, in effect, predictions). *The institutional imperatives (mores) derive from the goal and the methods.* The entire structure of technical and moral norms implements the final objective ... The mores of science possess a methodologic rationale but they are binding, not only because they are procedurally efficient, but because they are believed right and good. They are moral *as well as* technical prescriptions. (270; emphasis added)

When the norms are observed, scientific work will not be motivated by self-interest, knowledge will be evaluated critically and objectively without regard to the personal qualities or status of investigators, and scientific results will be open to the critical scrutiny of all. The disinterestedness prescribed by the norms merely reflects the universality and objectivity of scientific rationality.

Keeping the faith means acting rationally because the ethos of science prescribes norms of faith in scientific reason. Abiding by the ethos of science can become a condition of the possibility for the cognitive and epistemic value of scientific activity because to do so is to regulate one's own scientific endeavours and one's judgment of the work of others by purely epistemic criteria. The normative system of science is functional: it rewards work that passes the rigorous epistemic and cognitive standards of scientific method. The social features of Merton's system of scientific labour – its institutionalized normative structure – are therefore subordinated, as are Bacon's and Otlet's, to epistemological imperatives.

Given the imperfections of mortals, a system for distributing rewards for fostering a scientific conscience and adhering to the rigours of scientific method is needed. Truth is not its own reward because the production of scientific truth must demonstrate compliance with the ethos of science. But norm compliance is also not its own reward because scientists must be seen to comply and be recognized for it. The ethos of science is therefore supported by a system of adjudication in which scientific results are fairly judged and awards of merit bestowed upon those who genuinely advance knowledge. By contrast, Bacon's exclusive focus on the cognitive order of science failed to build norm compliance and its related system of adjudication and reward into scientific activity itself.

Merton's solution to the problem of how the reward system connects observing the ethos of science with production of work of genuinely cognitive and epistemic value is significant for a study of scientific documentation, because according to him norm-compliance, adjudication, and reward are relayed through the documentary circuits of the system of scientific information. The primary product of that system – the journal article – functions as the vehicle for the adjudication of norm compliance and distribution of reward. Merton holds that the most important kind of reward for scientific contributions is name recognition, or 'eponymy, the practice of affixing the name of the scientist to all or part of what he has found.' Examples of the most valuable rewards of this kind are designations such as 'Copernican system,' 'Hooke's Law,' and 'Haley's comet.' However, the 'large majority of scientists, like the large majority of artists, writers, doctors, bankers and bookkeepers, have little prospect of great and decisive originality. For most of us artisans of research, getting things into print becomes a symbolic equivalent to making a significant discovery' (1973e, 316). Thus for the great majority of scientists, journal publication distributes reward by bestowing some measure of eponymous recognition.

The value of the reward bestowed by publication increases with its visibility because, as Merton points out, 'for a published work to become a genuine contribution to science, it must, of course, be visible enough to be utilised by others' (1973a, 332). However, the very mechanism of publication – the referee system – is, as Merton puts it, an 'institutionalised pattern of evaluation.'[12] The favourable judgment of referees confers upon a document the imprimatur of science. Since claims for credit are honoured through publication in the journal literature, the cadre of editors and referees who manage the mechanism of journal publication administer one of science's most important reward systems. Merton says the 'referee is ... an example of status-judges who are charged with evaluating the quality of role-performance in a social system ... Status judges are integral to any system of social control through their evaluation of role-performance and their allocation of rewards for that performance' (Merton and Zuckerman 1973, 460). The mere publication of peer-reviewed work signals not only a unit of epistemic value, but also an investment in the further accumulation of value, since, as Merton shows in 'The Matthew Effect in Science,' reward usually generates further reward (1973b). Thus even though the majority of scientific publications are not used by anyone, mere publication itself is not without value.

The social system of reward supports the compatibility of norm compliance with disciplined adherence to science's cognitive standards. Rewards are meant to ensure that deviations from the technical and methodological imperatives of scientific work are at the same time violations of the norms of the ethos of science. But the organization of the reward system is embedded in the science information system. Whether scientists have violated either the methodological requirements of scientific rationality or the norms governing the scientific conscience is judged by referees who assess their formal writing, not by teams of inspectors who visit laboratories to assess the quality of experimental work. A reward system based upon document production supports the integration of cognitive and moral standards only if (1) the work of science consists in the production of abstract, epistemic content as grist for the mill of scientific rationality, that is, if scientific work consists in the production of 'thought and theory'; (2) peer-reviewed publications convey the appropriate epistemic content; and (3) the social system of peer review is properly disciplined to recognize the presence of valuable epistemic content in formal scientific writing. Given these assumptions, journal articles constitute work that meets science's cognitive standards, and publication constitutes reward for what is needed – the communication of scientific information – to further scientific knowledge production. Once scientific activity is identified with the production and communication of scientific information, then, with the norms in effect, the circulation of journal articles not only communicates the information required for the performance of advanced scientific work, as Bacon's model requires, but also distributes rewards to information of epistemic value, thereby satisfying Merton's model. In this respect, Merton's and Bacon's models are compatible. A reward system based upon citations serves as an example. If an article's references are traces of the information used to obtain its results – that is, traces of epistemic value – the comparative frequency of an article's citations becomes a measure of its epistemic value. Therefore a system in which citations count as reward is one conferring status upon the deserving – those who actually produce the epistemic content needed for constructing theoretical representations of the natural world.[13]

Merton has introduced, however unwittingly, a challenge to the privileged position of epistemology in scientific work. Since the reward for generating epistemic value for knowledge production is peer-reviewed journal article publication, it follows that the integration of the moral and the epistemic imperatives of science is achieved only if the journal article actually generates epistemic value for research science. If it does not, epistemic value becomes detached from the ethos of science. This could happen in at least two ways. First, the information required to derive further research results may in fact be communicated through channels other than science's formal literature. This creates a new problem. If the channels that do convey it operate beyond science's reward system, they escape its normative regulation. Second, it may be a mistake to assume a 'thought and theory' approach to the nature of scientific work. In the first case, the journal article becomes marginal to knowledge production, and the rewards of publication are detached from the ethos of science. In the second case, the significance of the journal article to the natural sciences is not measured by the information it contributes to the construction of conceptual fields. Pursuing the implications of the second alternative is taken up later in this book. The important point here is that when epistemological imperatives rule, the conclusion that the moral and cognitive imperatives of science are integrated by embedding its reward system for norm compliance in a system of production and circulation of documents requires the following assumptions: (1) science is a conceptual field of rationally ordered units of epistemic content; and (2) the primary role of documents is to convey this epistemic content. If it should turn out that the information conveyed by the journal article has little or no epistemic value for knowledge production at the research front, then from an epistemic perspective (discounting its epistemic value in gaining knowledge about science's social system), the journal article is, as it were, empty. In such a case, the reward system operating through journal publication regulates a field of epistemically worthless information.

The value of the social system of scientific information regulated by Merton's norms is therefore held hostage to the question, do scientific articles contribute information useful in the derivation of new results? Research into the scholarly communication among scientists suggests that they do not. Epistemological models of science's formal literature assume that they do.[14] In many places in his writing, Merton shows that he accepts such a model. Here is one, deploying the epistemological discourse of the 'reliability and cumulation of knowledge' through publication: 'the system of monitoring scientific work before it enters into the archives of science means that much of the time scientists can build upon the work of others with a degree of warranted confidence. It is in this sense that the structure of authority in science, in which the referee system occupies a central place, provides an institutional basis for the comparative reliability and cumulation of knowledge' (Merton and Zuckerman 1973, 495).[15]

The stress placed on the journal article by Merton's analysis introduces a destabilizing element into the status of the journal article in scientific work: the *possibility* of the reward system's continuing to function even if none other than referees or status-judges make any use of journal articles.[16] The rewards of publication (and even the rewards of citation) may be bestowed upon a scientific paper even if the value of its information for deriving new knowledge drops to zero. The Mertonian connection between epistemic value and norm compliance have been pried apart in precisely this way. Garvey (1979), for example, can assert that the information communicated in journal articles is not useful at the research front, where new knowledge is generated, yet its mere publication constitutes reward by establishing priority and ownership. In his and similar views, the article continues to communicate information, but not information of epistemic value for knowledge production. Rather, it communicates information used to secure social status for its author. A strength, albeit unintended, of Merton's model is its capacity to explain the importance of documents largely useless in the production of new scientific knowledge. But this explanatory power is gained at the expense of the communicative and informational value of the overt content of the scientific journal article, which played such a central role for Bacon. This overt, cognitive content is replaced by a latent content that helps secure rank for its author in a status hierarchy. Merton's norms therefore release a contaminating agent into the cognitive function of science's formal literature. Merton himself does not abandon his epistemological imperatives. But his insistence on social factors that can, in principle, drift far from the conceptual realm reveals the tensions between the two. The work required to secure their integration throws into stark relief the instabilities of his epistemological imperatives.

Science Information Systems: The Very Idea

Bacon's model of a science information system shows that the connections between science and its documentation have venerable historical origins. Both his and Otlet's insistence upon cognitive imperatives in regulating the systematic production, collection, and conceptual organization of scientific documents reveal the long ancestry and enduring significance of a key element of their analyses: the concept of the epistemic content of scientific documents, and of documents generally. Even when not made explicit, the assumption that the important thing about documents is their epistemic content – that is, their meaning, or the 'information' they either contain or somehow evoke in the mind of a reader – explains important recurrent features of how we think about systems of scientific communication. After all, what is a science information system supposed to do, other than communicate the meaning of the statements contained in its documents? Good information system design aims at ensuring the efficient movement of scientific information to the relevant sites of knowledge production. Indeed, the shift from document talk to information talk in library and information science, the academic discipline laying the most urgent claim to knowledge about science information systems and information systems generally, can be read as an expression of the notion that meanings, propositions, or epistemic content, whether existing somehow in the text or in the mind of a reader, are the kind of thing communicated via the ever-expanding variety of document forms and systems.[17]

Today, the talk is about information itself and its efficient communication, glossing over questions pertaining to its material form. These material forms – personal communications, conference presentations, journal articles, diagrams, charts, computer simulations, and many more, together with their institutional contexts – are important only because of their immaterial interiors: the information they convey. Even though he was far too sensible to abandon the material world in his analysis of scientific labour, Bacon nonetheless found in science's immaterial, cognitive order the blueprint for its supporting material, social organization.[18]

Our comfort with the notion of information itself comes as no surprise, given our saturation by statements in which information takes on the grammatical form of a substantive. It is presented as an object produced, transmitted, quantified, managed, bought and sold, needed and desired, an object whose speed of flow and global reach creates new social orders from ashes of the old: in short, it is an object about which vast new areas of knowledge are presented as opening before us. Epistemological models of science information systems remind us that talk about the immateriality of meanings, epistemic content, or propositions – that is, 'information itself' – supports talk about the immateriality of the activities in which this information is processed, manipulated, or otherwise implicated.

In Bacon's time there were fewer documentary options, hence no need to use a term like 'information' to generalize about a wide variety of documentary forms. So when he spoke of gathering scientific reports and accounts, he had in mind either writings his ideal documentalists discovered on their voyages, or the written reports they produced to record what was known but not already written down. In Bacon's case it would be anachronistic to call it 'information,' because the word's use to refer to what documents convey did not emerge until the mid-nineteenth century, as Nunberg (1996) has demonstrated. Bacon's 'information' referred to a process of becoming knowledgeable, as shown in this passage from his 'Preparative toward a Natural and Experimental History': 'As soon, therefore, as a history has been completed of all these things which I have mentioned – namely, generations, pretergenerations, arts, and experiments, it seems that nothing will remain unprovided whereby the sense can be equipped for *information* of the understanding. And then shall we be no longer kept dancing within little rings, like persons bewitched, but our range and circuit will be as wide as the compass of the world' (Bacon 1960, 277; emphasis added).

Yet in Bacon's time and in Otlet's, as in ours, the content of documents is the important thing, and for the same reason. Epistemic content is central to science when science itself is conceived as a conceptual field of organized knowledge, and when a dematerialized 'information' becomes the object manipulated by the correspondingly dematerialized conceptual and cognitive processes assumed to constitute science's paradigmatic activity: thinking. For both Bacon and Otlet, observation-propositions provide the inferential basis for higher-order 'axioms.' Theirs is a conception of science as an abstract entity, made up of dematerialized, conceptual substance, a conceptual field of propositions exhibiting more or less complexity of logical structure. Since the kind of substance communicated by documents – their epistemic content, or 'information' – is also dematerialized and abstract, it has the requisite ontological properties of raw material of science itself. Contemporary studies of science information systems would, therefore, seem to have gotten it right. The information is the important thing, not the material vehicle conveying it. Once epistemic content is communicated, the real work of science can begin, which consists in excogitating theoretical conclusions according to a specific and highly disciplined form of rationality whose historical origins reach back to the rules of scientific method that Bacon originated.

More than three centuries after Bacon, Merton found that science needed not only the epistemic content which for Bacon and Otlet regulated both the social system of science and the ideal state, but also the observance of a scientific faith. Recent studies of scientific information that have built upon Merton's ideas of credit and reward to find an alternative to epistemic content as the kind of information communicated by journal articles do not challenge the idea of information itself. Instead, after allowing that journal articles communicate information vital to establishing conclusions about science's social order, they simply look for the epistemic content valuable to knowledge production in other vehicles of the communication system of scientific information. Although some revisions admittedly may have to be made about what is communicated by specific documentary forms (the primary content of journal articles, for example, might be priority claims) information in the form of the epistemic content of the vehicles of scientific communication remains essential to science as long as science is conceived as a conceptual field.

Otlet's anxieties about the progress of the social sciences are grounded in the direct connection forged by the idea of epistemic content between science and its documents. If documents convey the epistemic content needed for scientific reasoning, inefficiencies in the system of communication translate directly into impediments to the progress of science itself. Others were less sanguine about the natural sciences, seeing them as overwhelmed by a flood of documents. Many shared Otlet's conviction that document reform would address the problem. Although Bernal's worries about the growth and quality of scientific publication were noted in the Introduction, the full extent of his commitments to the idea of information as epistemic content of documents are worth revisiting as a comparison with Otlet. Many of the inefficiencies of the communication system of science consist in obscuring what Bernal calls 'factual content' inside a larger unit, the journal article, that few have time to read and even fewer the capacity to comprehend. In his address to the 1958 International Conference on Scientific Information, Bernal recommends, in the spirit of Otlet, a new kind of publication – allowing for the exception of theories and ideas – which condenses the factual content of several related scientific papers into one document. He hopes that by 'passing all original material through the appropriate processes of classification and condensation, it would be possible to make at least their *factual content* also available to a much larger set of users' (1959, 88). The point of doing so is to permit the timely incorporation of the epistemic content scientists require for their current work. Under the present, chaotic system, there is a danger that valuable epistemic content is forever lost to scientific thought. Whether buried in journal articles, perspicuously presented in a reformed system of documentation, or before the mind of the working scientist in the form of an experimental result, only one kind of thing – information in the form of epistemic content pertinent to scientific reason – is present in each case. Bernal insisted that the 'working scientist or technologist needs information. He has the choice of getting that information through information services, or of finding it by experiment, or of working it out for himself' (79–80). The implication is that whether he finds it by experiment or works it out for himself, he acquires the same kind of thing that he gets through information services.

In his 1939 book, *The Social Function of Science*, Bernal remarked that because of the 'chaos of scientific publication' (119), '[m]ost of the scientist's work is in fact wasted ... there remains a considerable chance of its being entirely lost in the mess of unreadable scientific literature' (99). Waste was everywhere, but especially in the journal literature, since the 'number of scientific journals is altogether excessive' (118). The situation was sufficiently calamitous that piecemeal reform was out of the question: 'What is wanted is a ... drastic revision of the whole system of scientific communication' (119). Bernal's concern, expressed in the spirit of Otlet's emphasis on the documentation of 'facts,' stemmed from his realization that in the modern age the information made available through reports was more than ever the raw material of scientific production: 'As science grows, the facts on which it is founded and the way of building laws and theories from them depend less and less on the direct observation of nature by the scientific worker and more and more on the previous observations of other workers and on their methods of interpretation ... It is consequently of critical importance that the scientist at every stage in his work should be able to reach, rapidly and in convenient form, the results up to date of all relevant scientific knowledge' (117). His rough typology distinguishes between '(*a*) data to be used in practice or incorporated in research; (*b*) procedures, techniques and methods, including descriptions of apparatus; (*c*) conceptual frameworks, theories, or ideas' (1959, 82).

His remarks to the opening plenary session of the 1948 Royal Society Scientific Information Conference echo these themes: 'What is wrong with the present system is that the growing abundance of primary scientific publication and the confusion in which it is set out acts as a continuous brake, as an element of friction to the progress of science. We are not so much maintaining that scientific information is lost, though it may be, but that the scientific worker wastes time and effort in finding what information there is, and as a result we may be getting a far more limited and slower progress of research that we would under a better arranged system of publication' (Royal Society (Great Britain) 1948, 54).[19]

The epistemic concerns of Otlet and Bernal define the framework for research in science information systems. Its basic elements are generalized notions of *information* and *communication*. Although there are tensions between nominalist and realist uses of 'information,' discussed in chapter 2, the bias of a grammatical substantive is to suggest a singular sort of thing as its referent. However we might construe it theoretically, information itself would appear to be especially precious to the scientist. The thinkers canvassed in this chapter all hold that, within the circuits of socially regulated documentary systems, paradigmatic scientific activities consist in the methodologically disciplined pursuit and processing of the information required for the construction of scientific 'thought and theory.'

In the next chapter I trace some aspects of the troubled role of the concept of epistemic content as it appears in studies of science information systems in the literature of library and information science. This would be an easier task if the expression 'epistemic content' were actually used in such studies. However, since there is no shortage of uses of the term 'information,' and since epistemic content is simply the kind of information at stake in scientific documents, the burden shifts to tracing the referent of 'information.' This shift hardly renders the problem more tractable, however. Although 'information' is everywhere used as a grammatical substantive in such a way as to imply that the conclusions reached concern the referent of this term, explanations of what is used in studies of 'information uses,' what is needed in studies of 'information needs,' what is sought in studies of 'information seeking,' what is encountered in studies of 'information encountering,' or of that with respect to which one behaves in studies of 'information behaviour' are rare. Since the manifold conceptual confusion surrounding the term in the discipline has been definitively documented for some time by Alvin Schrader (1984a; 1984b; 1986), it would be futile to embark on a pursuit of its manifestation in studies of systems of scientific information and its use. I will pursue instead some traces of the notion of epistemic content in annual summaries of the year's best work in research on scholarly communication among scientists and in more recent, related work on information needs and uses generally.

2

Scientists and Other Information Users

This chapter reviews some of the epistemic narratives recounted in library and information science (LIS), the scholarly discipline with perhaps the highest stake in studies of science information systems. To manage a large literature in a small space, I rely first upon the annual reviews of the year's best research on science information systems and related areas from the *Annual Review of Information Science and Technology (ARIST)*. Little of the work reviewed in *ARIST* pertaining directly to science information systems is recent because, although studies of information use began with research on scientists, interest in other kinds of information users caught the discipline's attention from the mid-1980s (see Case 2002, 234). I then turn to more recent work on 'information behaviour' to show both the persistence of epistemic narratives of information and the tensions they exhibit between nominalist and realist interpretations of information. The chapter's final section, on the unhappy position of the beleaguered journal article in studies of scientific documentation, motivates my study in later chapters of the implications of a materialist alternative to these epistemic narratives, not only for the role of the journal article, but for documentation generally.

The *ARIST* Reviews of Information Needs and Uses

The first *ARIST* article on scientific information appeared in its inaugural volume in 1966. It described 1963 as 'something of a take-off point for empirical research on the information needs and uses of scientists and technologists' (Menzel 1966, 41). Its opening sentence states that the 'way in which scientists and engineers make use of the information systems at their disposal, the demands put to them, the satisfaction achieved by their efforts, and the resultant impact on their further work are among the items of knowledge which are necessary for the wise planning of science information systems and policy' (41). Two important factors set the stage for this 'take-off point.'

The first was the recognition of the need for and subsequent development of a state-sponsored system of scientific knowledge production. Vannevar Bush, one of the founders of the National Defense Research Committee in the United States and from 1941 director of the Office of Scientific Research and Development, was largely responsible for the development of this system, through which the production of scientific knowledge became vital to maintain the dominance achieved by the military in American society after the Second World War.[1] The scholarly study of scientific documentation was the academic response to the military significance of post-war science. In her 1978 ARIST review of research on information needs and uses, Susan Crawford writes: 'Use studies developed, in large part, as a response to the tremendous growth in science and technology subsequent to World War II. They originated with a practical objective based upon the notion that, if one could somehow identify the information needs and uses of a population subset, one could design effective information systems' (1978, 63). The militarization of research and development meant that efficiency became the primary objective of every component of scientific production. Science information systems especially cried out for reform, not only for the reasons expressed by critics such as Bernal, but also because prior to the mid-1940s their purposes were not shaped as much by military imperatives. Bush lamented the antiquated condition of the system of scientific information in his famous 1945 article As We May Think: 'The difficulty seems to be, not so much that we publish unduly in view of the extent and variety of present-day interests, but rather that publication has been extended far beyond our present ability to make real use of the record. The summation of human experience is being expanded at a prodigious rate, and the means we use for threading through the consequent maze to the momentarily important item is the same as we used in the days of square-rigged ships' (1991, 89). He proposed his now-famous 'memex,' a device often regarded as the precursor of hypertext because it permitted users to annotate texts and trace their own paths through a large store of documents made accessible from their desktops, as a technological solution to the problem.[2] Bush's vision of new ways to navigate masses of documents recognized the urgent need, imposed by the post-war mobilization of science for military research, for both vastly improved methods of information retrieval and novel approaches to its organization.

The second factor was the appropriation of 'information' in the late 1950s and early 1960s as the proper object of study by the academic discipline of library science. This development was contemporaneous with a proliferation of discourses of information in cybernetics, artificial intelligence, informatics, and literature, which Katherine Hayles describes so well in her 'story of how information lost its body' (1999, 2).[3] She shows that when information becomes an immaterial, theoretical entity, its disembodiment is articulated by diverse yet interrelated discourses of great complexity and cultural force. In remarking that 'a defining characteristic of the present cultural moment is the belief that information can circulate unchanged among different material substrates' (1), she captures library science's new idea. With science thoroughly integrated into a state-sponsored and policy-driven system of production, and with the meme of *disembodied information* set loose upon the culture, it comes as no surprise that the renamed discipline of library and *information science*, hungry for academic and scholarly legitimation and already claiming the study of scientific documents as its domain, should try to carve out a research area in science information systems – an area with significant policy implications.[4] Thus from the early 1960s, research on science information systems focused on engineering problems concerned with maximizing efficiencies in the flow of scientific information.

Realist Mansions, Nominalist Ghettos

Since its first volume, the *ARIST* reviews of information needs and uses have documented a tension between realist and nominalist uses of the term 'information.' Realist uses assume that the things denoted by the term share some real similarity by virtue of which it applies. Nominalist uses assume no common essence or real similarity between the items denoted by the term beyond their being called 'information'; the reasons for grouping them together are, from the perspective of the things themselves, arbitrary. Nominalism about information has the advantage of bringing together very loosely bounded sets of studies under a single term without worries about what information *itself* might be, usually in order to get on with solving the engineering problems of information system design. The tensions between these two uses are organized by the degree to which commitments to the idea of the epistemic content of documents are made explicit.

Realism about information is driven by a desire for the rigour promised by a robust conception of information as a unified, theoretical kind.[5] Such a conception defines the theoretical object of information studies. It satisfies the discipline's theoretical imagination and makes sense of the mass of information studies. If, for example, we do not assume that in selecting channel A over channel B, or information service C over service D, or document type E over type F, or message G over message H, a user pursues the same kind of thing, then these studies are not commensurable. That channel A is preferred over channel B is significant, and can provide a rational guide for systems design, only if users seek, and get, the same from each. If they seek, or get, something different from each channel, then channel preferences can be explained by differences in what users get, not in differences in the channels' capacities to deliver a singular kind of thing. Unless studies of information use assume that users actually use the same kind of thing – information – their channel comparisons become invalid. A fanciful example makes the point. If books are sought in order to decorate walls by shelf arrangements according to spine colour, and magazines are sought in order to balance uneven tables and chairs, these differences in 'channel' preference are explained by the differences in the uses to which the documents are put. It would be somewhat odd to call such document uses 'information uses.' The significance of channel studies that compare their communicative effectiveness depends upon the assumption that they convey the same kind of thing – just as studies of the efficiency of different kinds of hoses require tests with the same kinds of fluids flowing through them. In its realist sense, 'information' refers to the singular kind of thing conveyed by the many channels of communication. If it refers to a variety of things, no theoretical unification of disparate studies is achieved. Ben-Ami Lipetz makes the point this way: 'Before it becomes possible to design *from theory* an information system that will be ideal, or at least relatively effective, for the complex needs of a population of any size, we will require quantitative predictors of human needs and behaviour *with respect to information* that far surpasses, in detail and accuracy, anything available now or in the near future' (1970, 26; emphasis added). The possibility of designing information systems *from theory* rests upon a unified conception of information.

Realist tendencies surface in the many pleas for conceptual rigour emanating from gloomy prognostications of the discipline's intellectual poverty. Examples abound in the *ARIST* reviews. In the 1967 volume, Saul and Mary Herner plead self-defence for imposing their own organization upon the literature in face of a 'looseness or haziness of language [that] creates problems of understanding for reader and reviewer alike, and tends to defeat the purpose of publication, which is presumably to enlighten rather than confuse.' Their model of the communication of information, in which information itself emerges only at the final stage, is meant to introduce some rigour into such loose talk: 'The source or originator writes or speaks (method or activity) his ideas, research results, etc. (message), which are transmitted via a meeting, seminar, informational discussion, journal or primary publication (channel or medium) to the recipient who hears or reads (method or activity) the message and is thus informed. At this point, the message is converted into information' (1967, 3). This model is significant both for its commitment to a theoretically coherent concept of information and for its assumption that information is the epistemic content of documents. This is a model of a sequence of *activities*: first there is speaking or writing, followed by transmitting and receiving, then hearing or reading, and finally, becoming informed. Information appears at the end of the sequence: it is a state of mind, a mental condition, as it were, of the recipient of the message. Yet the model elides an intermediary step at the penultimate stage. It is not enough to hear or read a message to become informed. The recipient must also *understand* it, since hearing and reading are both possible without understanding.[6] But eliding understanding is not a fault of Herner and Herner's model. Rather, the implication that the unacknowledged assumption that a message is understood once it is heard or read is so obvious as not to merit mention demonstrates the power of epistemological conceptions of the 'information communication process.' It is only when the 'message' becomes present to the consciousness of the hearer or reader that it is 'converted into information.'

Even with the elision of a step generally taken for granted, Herner and Herner's model is a robustly explicit version of a realist conception of information as the content of the document present to minds that understand it. This conception is a leitmotif of many *ARIST* reviews. Thus in 1968 William J. Paisley, in what is otherwise perhaps the most urgent appeal among these reviews to recognize the many interrelated social systems in which the scientist is situated, nonetheless privileges in 'this regress of social and psychological systems ... *the scientist within his own head.'* He explains the latter 'system' as one 'of motivation, of intelligence and creativity, of cognitive structure, of perceived relevance of information inputs and uses of outputs.' This cognitive system is, he insists, the most important of all, since 'ultimately, all other systems support this one' (6). For Paisley, as for Herner and Herner, information comes into being as an object of consciousness in the form of the content of a message present to minds in a state of understanding.

The many reviews that criticize 'channel' or 'media' studies reinforce the cognitivist version of a realist conception of information that we find in Paisley and Herner and Herner. A typical complaint concerns the failure to distinguish between uses of channels and uses of the content they communicate (the channels studied include books, journals, abstracts, indexes, library catalogues, libraries, librarians, face-to-face encounters, conference proceedings and presentations, bibliographies, videotapes, films, government documents, research reports, preprints, reprints, personal filing systems, consultants, and databases; today the list includes electronic mail, the Internet, and the World Wide Web). Lin and Garvey (1972), for example, note that there have been far more studies of needs for channels than for 'substance' or 'the nature of the material.' Proposals for 'content' or 'substance' studies seek to make sense of the sheer quantity of information needs studies.[7] The idea is to get past the inessential, outer shell of information – the vehicle that communicates it – to the essential, inner core: the information itself. How is this core conceived? Thomas J. Allen complains that 'few studies have attempted to pursue information beyond its consumption by a user' to include 'the actual process of using the information' (1969, 5). This 'actual process' turns out to be a cognitive process. Allen places the few meritorious studies he has encountered in a category borrowed from Paisley ('the scientist within his own head') but renamed as 'the scientist or engineer as an information processor.' Examples of information processing are, appropriately, cognitive processes: *idea generation* and *problem definition*. The implication is that we have a deeper insight into channel use when we know the role of the content they convey in the cognitive processes of scientific thought. John Martyn's 1974 review makes a bold proposal: he would abandon the term 'information' altogether, because according to him, its use is chiefly responsible for leading information needs and uses studies away from their true object, the ideas communicated by various channels or media. 'Some readers may, with some justice,' he writes, 'conclude ... that information scientists are still studying how scientists use materials or services that contain ideas, and not how they use the ideas themselves ... The term "information" is the real handicap. If we abandoned the word "information" and thought, instead, of "communication," we might develop a more meaningful and productive mental attitude. Our proper preoccupation is, after all, with the interrelationship of people and ideas' (21). 'Ideas' are the important theoretical objects for information scientists; they are what scientists 'use' once received via the many channels of scientific communication.

Proposals to bring some theoretical rigour to information studies, whether focusing explicitly on the emergence of information itself in the mind of a recipient of a message or on 'content' rather than 'channels,' treat information as a theoretical kind – a kind of thing about which theoretical knowledge may be gained. Their cognitivist conception of information implies that the design of information systems can be modelled on principles of the cognitive processes involved in bringing representations of epistemic value to mind. In electronic information systems, for example, such principles are applied to ensure a smooth human-machine integration, or a fruitful 'human-computer interaction.'

When Herner and Herner remark that 'sometimes the word "information" is used to mean an end product (the answer to a question or acquired knowledge); other times it is used to mean media or conveyers of information such as journal articles' (3), they mean to highlight the conceptual confusion inhibiting a conception of information as a theoretical kind. But their complaint also reminds us that in many studies, 'information' is used in a nominalist sense. Rather than denoting a singular, unified, coherent kind that promises to provide a theoretical foundation for information studies, the atheoretical, nominalist use of 'information' gestures towards an ad hoc collection of studies unified only – but from a practical point of view, quite successfully – by imperatives of professional service to 'information users.' From a nominalist perspective, 'information' stands for anything that the community of information researchers accept as such to get on with the practical business of referring to a constantly growing set of document types (periodical information, book information, information in maps, encyclopedias, and manuals), media forms (electronic information, print information, digital information, information in films, videos, CD-ROMs, and DVDs), or content types (physics information, chemistry information, musical information). In such 'port-manteau' uses of the word, it has an open-ended denotation but no fixed connotation; it refers only to whatever document form, media type, or message content happens to be under investigation in a given case.[8]

The value of deploying a general term for a wide variety of cases consists in bringing loosely bound sets of research studies to the attention of readers sharing professional interests in improving specific services and systems of document acquisition, organization, and retrieval for specific categories of users. Such sets are unified by the managerial and engineering imperatives of an academic discipline and professional practice with roots in librarianship, not by a shared conception of information itself.

Although meant critically, Brenda Dervin and Michael Nilan's sketch of the 'systems-oriented' approach provides a useful summary of the nominalist orientations of channel studies: 'A typical study in the systems-oriented genre examines the extent to which a respondent (user or potential user of an information system) has: (1) used one or more information systems, used one or more different kinds of information services or materials; (2) sees one or more barriers to the use of the information system; and (3) reports satisfaction with various attributes of the system and access to it.' They go on to say that a 'typical study has tried to explain differences among respondents of these "information behaviour" dimensions with such predictors as demographic (e.g. age, education, sex), sociological (e.g. group membership), life style (e.g. interests and activities), and task description (e.g. purpose for contacting system)' (1986, 9–10). The generalizations spawned by such studies are similar to predictions of consumer behaviour based upon market research, where statistical regularities link consumption of specific products with consumers clustered by consumption behaviours.[9] Conclusions are supported by translating product consumption – the 'information uses' of information studies – into counts of instances of consumption and the statistical measures generated from them. Although yielding useful knowledge for improvements in information systems design as measured by user satisfaction, this kind of generalization about information uses employs 'information' to refer, in typically nominalist fashion, to a wide variety of channels, services, materials, and messages, without any explicit regard for what information itself might be.

As implied by those who complain that there are too many of them, channel studies provide good examples of nominalist uses of 'information.' Because many of these studies seek only to report that in a particular profession in a particular institution, performance of specific tasks reflect some measure of regularity in channel use, they have no need to provide theoretical reflection on the meaning of 'information' and its grammatical cousins, such as 'information needs' or 'information uses.' Unlike critics such as Herner and Herner, Paisley, and Allen, authors of channel studies do not have explicit theoretical commitments. Yet even nominalistically inclined studies embody traces of realism about information. Most if not all of the channel studies criticized as atheoretical in the *ARIST* reviews simply do not pursue what users do with documents once they retrieve them. If they did, there would be no reason for critics like Allen to plead for studies of 'the actual process of using the information.' The failure to do so suggests a deeply ingrained assumption that 'uses' of a channel consist simply in retrieving documents that convey useful messages, or yield 'user satisfaction' with the channel. Such an assumption explains why researchers feel no need to carry out the kind of investigations Allen proposes. Given that there are many uses of documents that have nothing to do with becoming informed by them (such as using them to level tables), the question left begging is, why does a *retrieval* count as a *use*? Presumably, because it is assumed that users read the document, understand its content, and become informed. Although not made explicit, the hidden assumption that information is that which is present to mind when a message is understood explains the conflation of retrieval and use. The tension between realist and nominalist uses of 'information' therefore maps the degree to which the idea of information as document content is made explicit. The fortunes of epistemic narratives in information needs and uses studies are traced by the strength and vigour of realist uses of 'information.'

The same tension also maps the distinction between the theoretical and professional motivations of library and information science. Engineering imperatives of information system design drives research towards study of a wide variety of 'information uses.' The desire to provide a unified theoretical grounding for the many disparate studies generated by engineering problems drives research towards conceptions of information as a theoretical kind. The distinction is acknowledged by Herner and Herner, who differentiate between basic and applied studies. Basic studies are those 'in which the purpose is to create or obtain knowledge regarding communication patterns and the reasons underlying them, with no particular application to specific programs or systems.' Their 'primary purpose is to create knowledge or understanding of the workings of man in relation to his information resources' (1967, 4). The assumption is that if we know, in general, the common thing sought in all searches for 'information,' then we have what we need to understand communication patterns per se, not merely the communication patterns of this or that specific organization or locale. The problem with applied studies is that they are too embedded in their locales; they may offer some guidance for practitioners working at the study site, but little or none to colleagues elsewhere. Consequently, such studies are quickly relegated to professional and non-theoretical areas of the LIS landscape. Thus Menzel, for example, excludes from his review those studies using 'categories too much tied to specific situations to make generalisations or extrapolations possible' (1966, 42). Allen makes the same point by remarking that he will 'exclude any study that does not have general implications beyond the need of a single organisation,' or those 'so involved with local circumstances that any generalisation is questionable' (1969, 3). Applied research also exhibits the limitations of studies based only upon uses of existing systems. Little more than minor adjustments to already flawed systems may be gained without a theoretical conception of the kind of thing scientists – or any other category of user – need and use. Bernal makes this point in his remarks to the Royal Society conference of 1958: 'The essential difficulty is that, though the user may well know what he *wants* from an information service, he is in no position to know what he *needs* from it, namely what variation in the system would help most to further his work. Consequently, any action based on analysis of present user habits is unlikely to produce impressive results' (79).

Bernal's point generates some anxieties for nominalist-inclined research on information uses. David Ellis's research devoted to behavioural studies of information retrieval (Ellis 1989; 1993; Ellis, Cox, and Hall 1993) broadens an insight from the first *ARIST* review, that study of scientific information needs and uses 'means primarily to study behaviour and experiences of scientists and technologists in confrontation with information channels' (Menzel 1966, 42). Refreshingly explicit about its method, Ellis's work challenges information retrieval engineers to demonstrate any need to displace search behaviours as the basis of retrieval system design. Ellis builds typologies of such behaviours by studying the interactions between existing systems and researchers in several disciplines. He concludes that robust categories of search behaviours, such as starting, chaining, browsing, extracting, monitoring, differentiating, verifying, and ending, are the most valuable guides for improving system performance. His investigations use 'information' simply as a handy 'port-manteau' word to generalize over an open-ended series of items sought and retrieved by the study subjects. The generalizations Ellis seeks do not derive from a notion of information as a unified, theoretical kind, but from the typicality of patterns of information-seeking behaviours. Whereas realist-inclined research speaks of a single kind of process (communication) and a single kind of processed object (information), behavioural approaches of a nominalist kind pursue solutions to engineering problems by appeal to behaviours that are, in some sense, typical.

A great strength of Ellis's behavioural approach is its rigour: its forceful articulation suggests that there is simply no need to investigate information as a theoretical kind to get on with the job of building better retrieval systems. Engineering problems are sufficient to provide valid criteria of significance for information needs and uses studies. One of the questions left begging, however (the one that exercised Bernal), is whether systems designed from theoretical conceptions of information itself can lead to greater engineering success than behavioural approaches. Studies of user behaviour with respect to existing systems have the virtue of applying their results directly, but they abandon Lipetz's vision of building an ideal system from a theory of the kind of thing users seek. Ellis's approach also raises problems of commensurability and comparability. Since his generalizations depend upon the typicality of user behaviour, there is no way of knowing whether they are artifacts of similarities in the design of existing systems. Moreover, how does one answer the question: typical of *what*? Of information behaviour? And what *kind* of behaviour is that? Engineering approaches can simply avoid such questions by recourse to improvements in systems design as measured by user satisfaction. As long as the behaviours studied yield categories that guide designers to more satisfying systems, they are, from a nominalist perspective, as robust and as legitimate as it is possible for them to be. If the aim is to maximize user satisfaction, then there is no need to concern oneself with what users do with items retrieved to their satisfaction, even if they are deployed to level tables or to decorate walls. That many would find such document uses anything but 'information uses' ought not to bother a rigorous nominalist, who can simply rule out of court anything but user satisfaction. Such nominalist rigour is, however, rarely on display. It is more plausible that lurking behind even those uses of 'information' that eschew any explicit attempt to come to grips with its referent is an assumption that what makes all this seeking exemplary of *information* seeking, or all these behaviours exemplary of *information* behaviour, is an implicit commitment to the idea that documents are *used* – and *used up* – when their content is present to users' minds. The tension between realism and nominalism about information, exhibited even in Ellis's research, is maintained by the degree to which such a commitment becomes explicit. Epistemic narratives are whispered beneath overtly nominalist narratives of information uses.

Although the dominant tendency of the *ARIST* reviews is to favour studies supporting the kinds of generalizations that might lead to scholarly research on information and communication per se, the very mention of the vast quantity of practitioner and engineering literature – the 'other' of library and information science's theoretical imagination – also invokes robust nominalist inclinations. Practitioner and engineering studies, seen as deficient because limited to highly specific institutions, categories of users, document forms, or channels of communication, live on the poor side of the tracks separating domains that seek a unified theory of information seeking from those that do not. Yet studies that carry no explicit implication of gaining knowledge about information in general, even though they may support generalizations at a level somewhat higher than 'the need of a single organization,' leave traces of realist tendencies. The mapping of the realist/nominalist distinction onto theoretical/practical and basic/applied distinctions functions to privilege epistemological conceptions of information as the epistemic content of documents as the preferred object of the discipline's theoretical desire.

The User as Hero

Although the tensions between nominalism and realism, and between practice and theory, were acknowledged in earlier *ARIST* reviews, they did not become the centre of attention until Dervin and Nilan's review of 1986. It identified 'a certain schizophrenia' at work in the information needs and uses literature since the previous review of 1978. 'On the one hand,' they write, 'the brunt of the work looks much like work reviewed in prior *ARIST* chapters' (9), that is, it is 'system-oriented.' However, what they claim to be new since 1978 is a heightened theoretical orientation, described as greater attention to 'identifying the underlying premises and assumptions' or the 'fundamental elements' of information needs and uses research. Chief among these fundamentals are 'definitions of information and need,' and 'the nature of information use' (12). In the terms I have suggested above, information emerges more fully as a *theoretical kind*. Even though the authors acknowledge that some of this new work supports 'traditional' approaches, it nonetheless helps advance our knowledge of information itself. 'Most observers agree,' they write, 'that information needs and uses studies have focussed on objective information, on a conception of information as something that has constant meaning and some element of absolute correspondence to reality.' Such assumptions 'have posited the user as a passive recipient of objective information,' or, even worse, people 'have been thought of as robotic information-processing systems' (13). They propose an 'alternative paradigm' which 'posits information as something constructed by human beings' (16).

From an epistemological perspective, however, a striking feature of their 'alternative paradigm' is the continuity between its conception of information as a theoretical kind and the 'objective' conception from which it wants to distance itself. I have argued that the most prevalent version of information as a theoretical kind (the kind of theoretical kind, as it were) is a cognitivist one, according to which information consists in the content of a document present to the mind of a user in a mental state of understanding that document. This view belongs to the idea of communication as the encoding of a message by a sender and its decoding by a recipient. Even though it is often the case, as Dervin and Nilan claim, that 'traditionally, research has focused on externals (e.g. contacts with sources and use of systems as indicators of need) rather than with internals (e.g. cognitive assessments)' (15), it is equally true that from the earliest *ARIST* reviews the clearest articulation of information as a theoretical kind was 'information in the head.' Dervin and Nilan's alternative paradigm points unequivocally in the same cognitivist direction. It speaks of shifts in the 'shapes of ... cognitive maps' (14); of 'the call for focusing on cognitive behaviour and the developing cognitive approaches to assessing information needs and uses' (15); of 'how people construct sense' (16) and 'when internal sense runs out' (17); of 'transforming image structures,' and of information as 'any stimulus that alters the cognitive structure of a receiver' (17); and of information needs as 'a conceptual incongruity in which the person's cognitive structure is not adequate to a task,' which takes the form of a person's recognition that 'something is wrong in his or her state of knowledge and wishes to resolve the anomaly,' or 'when the current state of possessed knowledge is less than needed' (17).

The difference between what Dervin and Nilan see as old and new paradigms of information is not one of information inside versus outside consciousness. In both cases, information is an object of consciousness. The difference is between two variants of a mentalistic conception of information. The 'traditional' cognitivism about information sees the mental process of bringing it to mind as less active and individualistic than the new cognitivism, which sees users as active beings 'who are free (within system constraints) to create from systems and situations whatever they choose' (16). They are not the 'input-output processors of information' of the old conception, in which communication is achieved when the information encoded by the sender is identical to the information decoded by the recipient. Yet in both cases, information arises as a result of the mental processes of an individual, whether they be the more active processes characteristic of 'sense-making beings' or the more passive (even 'robotic') processes of mere spectators of an objective world. Whether actively constructed or passively received, information achieves presence as an object of consciousness. Information is constructed inside the privacy of the mind, not outside, in the public realm, in shared social practices. For Dervin and Nilan, as much as for Paisley, the most important information system is 'the user inside his or her own head.'

In her 1978 review, Crawford remarked that 'the scope of use studies has been extended to include users in a wide variety of disciplines, among them psychology, education, policy making, and law. It appears that almost everyone's needs are now being surveyed – senior citizens, urban populations, minority groups, as well as scientists and technicians' (71). Some of the difference between the two flavours of mentalism may be due to this shift of attention away from scientists to a wider public. Scientists, after all, deal in facts, and the information they need consists in truths about the natural world. It would be disturbing to grant scientists *carte blanche* to 'create from situations and systems whatever they choose.' One might suppose that in such circumstances the system of scientific communication has broken down. Whereas we might be quite content with the notion that the information needs and uses of adolescent mall rats bear some relation to their 'construction of sense' by taking what they chose from the latest music video, we might be somewhat uneasy when information about the natural world emerges in like fashion in the minds of scientists. The shift away from the information needs and uses of scientists to 'almost everyone' can easily appear as a shift in 'paradigms.'

The individualist, mentalist, and cognitive emphases of Dervin and Nilan's new paradigm are nowhere more evident than in their response to the problem of 'chaotic individuality' that they see it as posing. If information is the result of a radically individualistic, private, mental, sense making process, how can any systematic connection be found between the sense made of the same message communicated to different individuals? Indeed, what would the 'same' message even mean in the case of individual sense making? Their solution is to resort to 'the fundamentals of the human condition as a means for systematising individuality' (15–16), and to search for 'universal dimensions of sense making' (16). Since these fundamentals and universals must be cashed out as mental processes, it is an open question whether, once discovered, they render the information user significantly less 'robotic' than the information processors of the old paradigm. Moreover, awaiting discovery of the fundamentals of the human condition and universal dimensions of sense making might be thought by sympathizers of science's apparently lower threshold of tolerance for highly individualistic and creative acts of sense making to unduly delay our understanding of science information systems.

Epistemic Narratives of User Studies

The idea of information as the epistemic content of documents, which I claim is a concept that appears throughout the *ARIST* reviews, reappears in various forms and to various degrees in other user studies. In some of them, this idea intersects with concerns about the information needs and uses of scientists. But the degree to which the documentary approach to information studies developed in this book may be generalized beyond scientific documentation depends upon assessing the role of this idea in studies of information seeking and information behaviour generally. Four types of studies are addressed here; two that focus on scientists, and two of broader scope.

Typologies of Content

Several of the *ARIST* reviews call for closer attention to different kinds of scientific research tasks, so that sharper distinctions between information needs and uses might be revealed. This recommendation is based on the conviction that in 'the basic sciences, the principal factor influencing the scientist's needs and uses of information is his association with a research area' (Crane 1971, 4). With a clearer understanding of the scientist's research tasks, we should be able to 'glimpse a real scientist or technologist at work, under constraints and pressures, creating products, drawing upon the elaborate communication network that connects him with sources of necessary knowledge' (Paisley 1968, 2). How have these research tasks been conceived? In what sorts of projects is the scientist thought to be occupied?

The first of my four cases is Menzel's effort to expand the typology of 'science-information functions' beyond the then-pervasive dichotomy of 'reference' and 'current awareness' to include three others: exhaustive search, 'brushing-up' (satisfying scientists' efforts to familiarize themselves with fields of interest new to them), and the 'stimulation' function ('stimulating researchers ... to seek information outside their predefined areas of attention,' beyond 'the informational requirements each scientist can define for himself') (1964, 9–11). His expanded typology is based on the insight that research scientists do many things falling under the category of 'using information.' Questions about 'information needs' without further differentiation gloss over the very distinctions among needs and uses that should be central to the design of effective science information systems. His advice to researchers reminds us that 'information need' is an LIS term of art, not one readily intelligible to anyone outside the field: *'the information needs of scientists cannot be ascertained simply by asking a cross-section of scientists, "What are your information needs"* let alone by asking them, "at should be done to satisfy your information needs"' (Menzel 1964, 15; emphasis in original).

Menzel's work merits attention forty years after its publication, not because of an abiding conviction in the validity of his typology but because, in spite of his salutary objective of enlarging the scope of 'information uses,' it reveals the epistemological assumptions driving his conception of the scientific enterprise. His categories of information functions are based on the different kinds of content communicated by documents: new developments in a field (currency); specific data (reference); comprehensive subject coverage (exhaustive search); subject areas related to specific research fields ('brushing-up' and 'stimulation'). *Content* or *subject matter* is Menzel's principle of research task differentiation. His mapping of differences in the information uses of scientists onto differences in document content revisits the assumption noted earlier: that an 'information use' consists in the *comprehension* of a message conveyed through the channels of scientific communication – a text is *used* when it is *understood*. Although Menzel does not talk about information as presence to the mind of a reader of a text, his derivation of a typology of information uses from differences in document content, whatever the document might be – a conversation with a colleague, a preprint, a book, a conference presentation, a journal article, even observation of nature itself – implies the mentalist assumption. Information is once again imagined as a theoretical kind: something stable, uniform, and wholly present. Differences in information uses consist only in what is represented – in what the information is *about*.

The idea of information as the epistemic content of documents links information systems to systems of knowledge production. Information is imagined as the kind of thing scientists work with in their privileged sphere of activity: *inside their own heads*. To ask, of a specific text, exactly how it is related to the complexities of scientific work becomes a question about how a specific unit of epistemic content, derived from the text and now present to the mind of the scientist, is connected to other units of epistemic content by the application of scientific reason. The derivation of a new result from such connections is the aim of scientific knowledge production.

The relays between information and the epistemic content of documents also distinguish the professional jurisdictions of the scientist from those of the information specialist. The question of the relations between units of epistemic content belongs to the scientist's own knowledge domain. It is not within the competence of the nonscientist. It is a question about expertise in the exercise of the highly disciplined and specialized activity of scientific reasoning. If the role of an information system is to be as responsive as possible to the 'content' needs of scientists, then it performs its service by delivering documents that are *used* – that is, read and comprehended. How their content figures in the scientist's thinking is not, and cannot be, the concern of the information system or information providers. The limits of the science information system in its relationship to scientific research is described by Mikhailov, Chernyi, and Giliarevski as fundamental: 'The scientific information activity ... strives to raise the effectiveness of research and development and consists of the selection, analytical/synthetic processing, storage, and retrieval of scientific information contained in documents and the presentation of this information to scientists, researchers, and specialists in a timely manner and in a form convenient to them' (1984, 226). Arguing against 'the attempt to posit the equal sign between research and the scientific information activity' (227), the authors explicitly differentiate LIS work from scientific research itself: 'In contrast to research proper, the scientific information activity is the first, preparatory stage of any research. It creates conditions for the more effective conduct of research ...' (228). Menzel's focus on content as the defining principle of research task differentiation suggests the same limitation. Its rationale lies in the assumption that the fundamental scientific activity is thinking – a private, mental activity. It follows that the best that even an optimal system can do is to maximize the number of satisfactory documents scientists retrieve in seeking the 'content' only they know how to use in the production of new results.

Typologies of Method

My second example consists in attempts to map 'steps in the research process' to 'probable uses of recorded knowledge.' A typical formulation of the methodological steps research scientists are thought to take is the following: '(1) perception of a problem; (2) definition, or precise statement of the problem; (3) formulation of the hypothesis; (4) choice of method; (5) choice of data gathering and analyzing techniques; (6) search for evidence or data; (7) generation of conclusions and corollary implications; (8) discussion of consequences' (Egan and Henkle 1956, 142). The authors then map specific information uses to each of these 'steps in the research process' in a list under the heading 'Probable Use of Recorded Knowledge.' For example, the first is alleged to correspond to uses of 'theoretical treatises, or research reports,' the second to 'intensive search for prior investigations of the same problem or of similar problems,' and 'search for accepted definitions of relevant concepts,' and so on, for each of the eight steps.

Margaret Egan and Herman H. Henkle's model rewards study because it is a familiar type.[10] It features an uncritical acceptance of an idealized conception of scientific method rooted in stages of *thinking* directed towards the production of *propositions*. The behavioural study of scientific information use and exchange conducted by Garvey (1979) and his collaborators follows a similar logic. They are concerned with charting processes of production, dissemination, and use of various forms of scientific information, whether communicated through formal or informal channels. We learn much about the use and exchange of prepublication reports, oral reports at conferences and national meetings, and preprints; of submitting, vetting, and accepting article manuscripts, the relationship between journal use and previous awareness of the content of articles through informal channels, and the connections between articles and science's secondary literature. We also learn about the links between uses of various kinds of information sources and the different stages of scientific activity. But like Egan and Henkle's model, this work is based on idealizations of scientific method. Garvey says, for example: 'The sequence of stages displayed by these results is much like that taught science students and the experienced scientist may not consciously follow the sequence rigorously. But the important thing for information services is that a rational ordering exists which can be of advantage' (260–1). Egan and Henckle demonstrate a readiness to accept conflicts between their proposed and actual use of specific materials at specific stages of scientific work. The authors seek to placate any anxiety occasioned in the reader by the use of 'probable' in their heading by commenting: 'Our knowledge that the materials mentioned above in association with successive steps in the research pattern are actually so used depends not only upon the logic of the association but also upon considerable pragmatic experience and observation, and upon a few systematic studies.' The attractiveness of the model derives from the apparent 'logic' of its steps. Their concession, that 'such use is neither invariable or efficient' shows that their model owes more to allegiances to an idealized 'scientific method' than to actual practice. Alluding to one of the assumed 'uses of recorded knowledge' – the 'intensive search for prior investigations' – the authors observe: 'There is good reason to believe, for instance, that a great deal of both basic and developmental research is undertaken with only a cursory examination of the record' (142–3).[11] As for efficiency, the matter is even worse: 'As to whether or not the materials mentioned in the other steps are used in a satisfactory way there is little evidence' (143).

Like Garvey, Egan and Henckle admit that they derive their notion of scientific method from its representation in the literature of science: 'Common to research workers is a pattern of thought which, having been thoroughly tested and found valid, is formalised and prescribed for use in all research investigations. Although the considerable body of literature on scientific methodology includes descriptions and justifications of numerous variations upon the accepted method of procedure, there runs throughout a basic pattern which is so well-established, and so well-suited to a variety of problems and fields, that it is taught to all students who are embarking upon research' (138). But it is a risky business to rely, in one's specification of research tasks, on textbook versions of a formalized and prescribed 'pattern of thought' designed for students. A major theme of chapter 5 is the distance between the literature's reconstruction of an ideal scientific method and what actually happens in laboratory practice. At this point, I only draw attention to how the activities enumerated in Egan and Henkle's model are conceived. When scientific work is imagined epistemologically, it comes as no surprise that formal reconstructions of scientific method privilege patterns of *thought*. The model cites perceiving, defining, formulating hypotheses, deciding on methods and techniques, and generating conclusions – all cognitive processes belonging to an idealized, 'thought and theory' version of scientific rationality. The scientist is envisioned as a thinking creature engaged in the mental processing of the epistemic content of the documents 'used' at each stage of a methodical scientific process. This view privileges the seeking and processing of information in the research process, and it installs the communication of information at the centre of scientific knowledge production. But the studies of scientific practices discussed in later chapters show that the method featured in the literature is a fiction, constructed post hoc, bearing a stronger relation to the rhetorical imperatives of writing scientific papers than to realities of scientific labour. A fantasized, representational, knowledge-and-information model of scientific activity can only map onto a fabled terrain of information uses. Readers are left to imagine what a new and improved science might look like whose practitioners fully exploited the knowledge possessed by information scientists about the information uses corresponding to each step of scientific method.[12]

As later chapters show, recent studies of scientific practices have shifted the focus of science studies from this overly philosophical conception of scientific labour as the production of conceptual fields to material practices of 'tinkering' and getting things to work. This shift poses a new problem for the role of documents in scientific labour because it challenges the centrality of information and its associated cognitive processing in scientific practices. This new problem is best approached by showing how documents are situated within science's material practices. In the work of Egan and Henckle, Garvey, and much other research on the information needs and uses of scientists, the 'real scientist or technologist at work,' as Paisley put it, remains an elusive figure.

World Three Enthusiasms

My third example shows that a realist conception of information as epistemic content, and thus as a theoretical kind, is not limited to an idealist ontology in which information exists only as an object of consciousness. Although its idealist interpretation is explicit in studies of information use that are biased towards cognitivistic and mentalistic interpretations and implicit in many others, conceptions of information as existing beyond consciousness are also found in the LIS literature. They are realist in a double sense, in contrast to both nominalist and idealist conceptions. The latter contrast turns on whether one holds, with idealists, that the objects of our knowledge are no more than our own representations. Cognitive conceptions of information are realist because they view information as a particular kind of thing, but idealist by virtue of construing it as an object of consciousness. To use 'information' in a doubly realist sense is to hold that information is both a theoretical kind and exists beyond our consciousness.

The most explicit version of this doubly realist view emerged in the 1980s among a small group of enthusiasts who championed Karl Popper's notion of World Three as a new domain for LIS (Brookes 1980a; Davies 1989; Neill 1982; 1987; Swanson 1986; 1987; 1989a; 1989b; 1990a; 1990b). This group interprets scientific activity as explorations of World Three, 'the world of objective knowledge – the world of problems, theories, and other products of the human mind,' according to Don Swanson's (1986, 107) gloss of Popper. The reference to the human mind should not suggest that World Three consists of mental substance; that would be to confuse it with World Two, 'the world of subjective knowledge or experience, of mental states and mental processes.' World Three, although a human creation, 'contains far more than man has ever thought of or dreamed about'; it is populated by abstract entities, such as meanings and propositions, including all of the implications of every proposition. Swanson gives an example: 'man created the number system, but once it was invented, unintended and unforeseen consequences, including prime numbers, then followed – and awaited discovery' (107).

Champions of World Three see an opening for the direct participation of information professionals in the scientific process. Swanson's hypotheses that dietary fish oil ameliorates Reynaud's disease, and that magnesium deficiency is a cause of migraine, are generated from computer searches of the scientific literature. Roy Davies sees document classification and information retrieval as methods of scientific knowledge generation. Sam Neill sees World Three as terrain that can and ought to be claimed for many LIS research areas. Information professionals can be thought to produce scientific hypotheses through information retrieval, document classification, and online searching because scientific work is imagined as the production and processing of ideas, propositions, and meanings. When information is imagined as the content of documents, information scientists are interpreted as handling the same kinds of things as natural scientists. Their methods are different, because the former derive their techniques from information science. But the result can be the same, as Swanson insists: the generation of hypotheses from existing information. Once produced by scientific thought, 'information' becomes a real object existing not in the mind, but in World Three, a world hospitable to colonization by information scientists. A doubly realist, epistemological conception of information is in full flower here. As imagined by World Three enthusiasts, it legitimates displacing the natural scientist's assumed exclusive rights to scientific statements, and violates the injunction against 'the attempt to posit the equal sign between research and the scientific information activity' issued by Mikhailov, Chernyi, and Giliarevski. In World Three, information scientists become the Lamps of Bacon's utopia.

From Information Seeking to Information Behaviour

My fourth example is a set of recent studies of information seeking, information behaviour, and information practices. These may be seen as an acceleration of the trend, noted by Crawford in 1978 (see p. 66), of extending user studies to a wide variety of populations. The study of information seeking beyond professional and academic settings is now sufficiently mature to enjoy its own designation: 'everyday life information seeking' (ELIS). Studies in this area conducted in the past quarter of a century, and especially after Dervin's work made itself felt in LIS, are distinguished by their interest in cases of information acquisition by 'ordinary people' in their daily routines. Recent work in this vein broadens information seeking to include the 'information encounter,' or the serendipitous acquisition of information – a development whose significance is perhaps best measured not by the novelty of its key idea, that we often learn things unexpectedly, which is familiar enough, but instead by the degree to which it has moved beyond information retrieval research, the precursor of information seeking studies. 'Information encountering' sparked further developments. The concepts of information seeking and information behaviour were found to be limiting because of their connotations of acquiring information as a response to a well-defined problem. Information encountering was also seen as unduly narrow because it assumes a clear distinction between 'active' and 'incidental' information practices.[13] Themes already encountered in my previous three examples are found in this area of study as well: tensions between realist and nominalist uses of 'information,' different versions of the idea of information as the epistemic content of documents, and the possibility of relating disparate kinds of studies by means of a coherent articulation of their key concept: information.

Pamela J. McKenzie's work is a good place to begin, for several reasons. First, it evidences a rare acknowledgment of the distinction between the study subjects' information practices and their accounts of them. Many studies rely on such accounts, but treat them as unproblematic, wholly veridical descriptions. McKenzie does not; she says: 'The model presented here [of the information practices of nineteen women pregnant with twins] ... is derived from accounts of information seeking and not from observation of information seeking as it happened' (2003, 22). She makes it clear that her research is about the stories her subjects tell about seeking and encountering information. Second, the stories themselves are rich and textured, moving fluidly from one life-scene to another, easily transgressing, as McKenzie points out along the way, the sharp distinctions often drawn between types of information seeking and encountering. Third, her two-dimensional model captures a wide spectrum of practices, ranging from actively seeking out answers to specific questions to receptively placing oneself in an information-rich site where valuable information can accidentally and serendipitously be gained.[14]

What kinds of stories are these? For the most part, they are first-person stories, told by women who, whether by overt design, by explicit attention to relevant parts of their environment, or by merely being receptive, wish to put themselves into situations of *coming to know* as much as possible about being pregnant with, giving birth to, and raising twins. Insofar as they are stories about acquiring knowledge, they are epistemic narratives; examples include calling a doctor or a friend, seeking a specific book title, receiving unsolicited magazine articles from relatives, browsing bookstores, reading the newspaper, and so on. Documents play a role only in some of these stories, for example, in seeking a particular book, accidentally encountering a relevant newspaper article, or simply browsing in a bookstore. Other cases involve conversations, whether engaged in serendipitously or with intent. Some cases involve neither, such as an instance where a subject came to realize the disadvantages of the side-by-side baby carriage she favoured upon watching a father with twins trying to manœvre one through a narrow department store check-out counter (34).

The richness and variety of McKenzie's cases have the advantage of expanding the research field. But they also pose a problem encountered in previous examples. Their disparity suggests a robust nominalism about information in calling them all 'information practices.' As we have seen, a theoretical unification of information studies, where individual studies can be brought together to bear on shared theoretical issues, is a function of the strength of allegiances to realism about information. The stories McKenzie's subjects have to tell are rich and textured, but the stories of interest to me are those LIS researchers tell. By presenting her examples as instances of a particular kind of activity – 'information practices' – McKenzie constructs her own disciplinary and scholarly narrative. Its coherence derives at least partly from its epistemic features, which draw upon a cultural habit of imagining the process of coming to know as attaining a particular *state of mind*. When instances of coming to know by reading documents are placed side by side with those of acquiring knowledge through observation, as in the case of the baby carriage at the check-out counter, a narrative that presents them as members of the same set gains credibility from the idea that in each, a message is sent and received: sometimes from documents, sometimes from the world. Nunberg's 'intentional substance present in the world' – the *phenomenon* of information as it presents itself to us today, a phenomenon of epistemic content or 'information,' liberated from documents and inhabiting the world as a pure essence – underwrites, especially in the absence of any explicit theoretical treatment of what 'information practices' might be, the power of McKenzie's narrative to bring these disparate *comings to know* together in one coherent story. Our acceptance of this wide variety of ways of gaining knowledge as instantiations of the category of 'information practices' depends upon three elements: an explicit *absence* of theoretical concern for explicating what 'information' practices might be; the cultural phenomena of information and knowledge as a unique, theoretical kind; and the cultural form of contemporary academic discourse, which consists in weaving many different research strands together in a narrative of a continuing collective venture in which individual studies gain value by pointing in new research directions.[15]

The narrative coherence of McKenzie's work therefore depends upon its implied statement that the cases studied are all instances of 'information practices.' In J.L. Austin's terms, this is not a constative assertion (it does not make a statement about the world that is true or false) but rather a performative one, which might be rendered in this way: 'I hereby invite you to consider these cases as instances of a single category, "information practices," in the interest of sowing the seeds of fruitful new research in information studies.' The risk one runs, of course, is that others will not take the indicated path, but a worse alternative may be that the proffered licence to treat even cases of gaining knowledge by simply observing events as 'information practices' may be so enthusiastically received as to fragment the field of 'ordinary life information practices' beyond the hope of establishing useful and theoretically productive relationships between them. The grammatical connections between information, becoming informed, and gaining knowledge work together to subvert McKenzie's scholarly ecumenicalism about information practices. Deflating information alleviates the discomforts of tensions between nominalist and realist proclivities.

The work of Karen E. Pettigrew presents a contrast to McKenzie's regarding concessions to theoretical exigencies. Where McKenzie rests content with no definition of information, Pettigrew's struggles with the issue evidence a rare acknowledgment that some conception of the nature of information is at stake. 'Defining the concept "information,"' she writes, 'has been a continuing and much-debated topic in LIS and it is increasingly apparent that there is no single definition of information that will work in every research setting since information takes on different meanings in different contexts. Thus, when operationalizing "information," researchers should consider the paradigm from which they are working and the nature of the specific setting in which they are studying information' (1999, 809). The claim that *information* (rather than 'information') takes on different meanings in different contexts suggests *something* that dons and sheds meanings, perhaps much like an anxious adolescent who, prior to a first date, changes outfits before the mirror – a reasonably robust realist conception, hence one closely aligned with ideas of epistemic content and disembodied messages. But the question here is whether 'information' – the term – takes on different meanings in different contexts. If it does, and if, as appears from Pettigrew's later remarks, this is really the issue that interests her, then her position exhibits nominalist tendencies as well. Yet longings for definitions evidence realist urges, since the point of providing them is to reach some conceptual clarity about the *kind* of thing, process, state, or event defined. The tensions at work here recognize both the specificity of Pettigrew's case study – she tells stories of conversations between seniors and health care providers at foot clinics – and disciplinary obligations to situate her narrative in an ongoing, collective research program which is the source of its theoretical legitimation.

These same tensions are evident in her operationalized and contextual definition of 'information.' At first glance it seems radically specific, with no obvious traces of the generality such definitions usually display. True to its context, it is not a definition of information itself, but of what it means in her particular research setting. She therefore defines 'human services information' (HSI) as follows: 'a communicative construct involving the nature or availability of local services and programs that is produced in a social community-based context' (811).

Although her definition is intended to show 'how information, HSI specifically, could be defined in the context of this research' (809), a closer look reveals that it abstracts from that context in several respects. The contextual features are: (1) there is a conversation between at least two people, at least one a senior and at least one a health care provider; (2) whoever gives HSI believes that it is capable of effecting a cognitive change in the recipient's view of his or her situation or problem; (3) a social interaction occurs between the parties to the conversation, in which the discussion of HSI is embedded; HSI is neither the first nor the only content of the conversation; (4) HSI is not 'morselized' or bounded; it develops in conversation, often in questioning and commenting; (5) HSI is embedded in a complex interaction, in the sense that the parties to the conversation have different strategies and degrees of interest in what is said; and (6) the conversation has effects other than simply communicating the literal content of HSI (e.g., clarifying a situation, suggesting ideas, providing affective benefits) – a feature described as the 'multiplex' nature of HSI (810).

It would be clumsy to incorporate all six of these aspects of HSI in one definition, but if one did, in the interests of more rigorous attention to context, the abstractions of Pettigew's definition become clearer. Consider one that incorporates just some of these aspects. For example: 'HSI is communication achieved through conversations in foot clinics between seniors and health care providers about health services, in which the provider believes that what is communicated can effect a cognitive change in the recipient's view of his or her situation, and including any and all cognitive, social, affective, and instrumental benefits the seniors involved in the conversation may gain from its use.' A definition that captured all six of Pettigrew's essential features of HSI would have the consequence that no HSI is conveyed if (1) it is communicated at sites other than foot clinics; *or* (2) it is conveyed other than through conversation; *or* (3) the person giving the HSI did not believe it could effect a cognitive change in the recipient's situation; *or* (4) it is conveyed immediately rather than 'mutually constructed' in conversation; *or* (5) the parties to the conversation do not have different strategies or degrees of interest; *or* (6) it communicates nothing but its literal content and is not used for any other benefit. While true to the context, as Pettigrew says a good definition of information should be, this definition is so wedded to its context that HSI likely belongs to a category of one. Any transgression of even one of the six elements would imply – by definition – that whatever is studied is not HSI.

Pettigrew's definition of HSI abstracts significantly from contextual features. How true to context should a definition of 'information' be? This question poses with some urgency the problem of the commensurability of her studies with others. The problem is not an artifact of a definition extended to include all six features of HSI. Pettigrew's definition follows Kimmo Tuominen and Reijo Savolainen (1997) in restricting the use of 'information' to what is 'constructed' through conversation. Many studies of 'ordinary life information practices' do not involve conversations. For example, not all of McKenzie's cases do. But if the meaning of 'information' in such studies is different from its meaning in Pettigrew's work, then so is the meaning of its grammatical cousins: 'information seeking,' 'information encountering,' and 'information practices.' We are faced once again with the fragmentation of information studies: such definitional variations incur because they imply that different phenomena are present in different contexts.

Pettigrew's fidelity to contextual definitions of information do not prevent her from seeking to make her work useful to others. Her own definition situates it in the research area described by Tuominen and

Savolainen. She also observes that 'researchers construct theories of IB [information behaviour] that may be used as frameworks for studying other populations or settings and for deriving questions for further invesitgation' (802). 'Future research,' she explains, 'might explore whether the flow of human services information at diabetes or wellness clinics, hair salons, quilting bees, etc., is affected by similar contextual factors and if information might be defined using a similar social constructionist conceptualisation' (814). There is clearly a great deal of work yet to be done. The researching subject projected by her work is the team player, dedicated to the success of a shared project, in which researchers with interests in many regions of the information-seeking landscape submit contextually rich studies like her own in the interests of a collective gain in empirical and theoretical knowledge of information behaviour. Yet the strength of the bonds between her work and others' consists, as in McKenzie's case, more in what is not made explicit than what is. Narrative extensions to other research contexts are constructed by *undefined* talk about information flows, becoming informed, sharing information, and those who 'give' information and those who 'receive' it; messages conveying a particular kind of content (HSI) are sent and received. There are interesting differences between Pettigrew's and McKenzie's work, to be sure. Pettigrew restricts the flow of information to conversations, while McKenzie allows individual observation of events of the world; and whereas McKenzie restricts her instances of information acquisition to cases of coming to know, Pettigrew allows that the recipient may gain no knowledge – perhaps one has heard that HSI before – but may nonetheless gain some affective benefit simply from hearing it in this caring context, or may put it to some instrumental use later on – perhaps gaining some advantage from the fact that it was conveyed by a professional at a clinic. McKenzie's narrative is always epistemic, whereas Pettigrew's allows that information is implicated not only in knowledge (cognition) but also in emotion (affect) and motives (instrumentality). Yet in both cases, the language of information underwrites implicit proposals for extending the reported research to other contexts because of the realism about information embedded in its discourse of flows, transmission, sending and receiving, and the contents of messages. The strength of connections between Pettigrew's narrative and other work depends primarily upon discursive invocations of the trope of information as intentional substance flowing through conversations and documents, and from the eyes monitoring the world to the mind behind. Ironically, explicit appeals to contextual definitions work against such connections, even granting their abstractions from context, due to the 'stickiness' of the local context their rich descriptions convey.

McKenzie's and Pettigrew's work show that nominalist tendencies towards fragmentation of information studies are checked by realist counter-tendencies that owe their strength to uses of 'information' which draw upon different strands of the cultural phenomenon of information as a theoretically coherent kind. Many information studies can be positioned on a scale of strength of commitments to realism and nominalism about information. The work of Catherine Ross on reading for pleasure (1999) lies at the extremes of weak realism and strong nominalism. Ross recounts stories of pleasure reading as told by 'heavy readers'; she finds that because they integrate reading into the texture of life, the books they read have a significant impact upon them, revealing new perspectives or enlargements of possibilities; they provide models for identity, reassurance, comfort, confirmation of self-worth, the acquisition of strength or courage to make life changes, and acceptance of others (793–5). The examples are powerful and highly interesting, and they tell us much about pleasure reading. But Ross wants them to tell us something about encountering information. This is surprising, on at least two counts.

First, it is unlikely that one would be inclined to designate 'information encounters' those awakenings, gainings of courage, and so on, inspired by say, a beautiful sunset, a sympathetic and loving friend, the death of a loved one, or acts of courage and mercy. This is particularly true if the emotion involved is especially strong; if I gain enlightenment after many hard years of studying Zen Buddhism, and as a result my life is changed, I would find it trivializing to interpret my experience as an 'information encounter.'

Second, Ross's use of the term 'information' is highly selective. She rarely employs it in her compelling descriptions of the role pleasure reading plays in the life of her study subjects. There are only three such uses in her interview reports. The first appears in a case of pleasure reading that is *not* about making a difference in life, but about learning a skill: one of her subjects reports reading about embroidery or gardening before undertaking such activities (788). The second occurs in a speculation that how readers find books for pleasure reading may be 'extensions and adaptations of everyday practices that they typically find useful in information seeking' – an application of LIS terms of art to make a connection between her study and *other* kinds of cases, but not a description of the case itself. The third appears in a report of an interviewee's comments about how books help with problems in her life even though she does not explicitly seek books to match the problems. By interpreting this encounter with helpful books as yielding 'answers to current concerns,' Ross uses language that invokes the problem-solving approach of traditional information-seeking studies, yet she simultaneously distances her own work from them and locates it in the 'information encountering' arena by describing the way her study subject found helpful books as a 'process ... without the need for active information-seeking' (795). 'Information' was not used in *any* of the other descriptions of cases of reading for pleasure. The great majority of uses of 'information,' either on its own or in canonical disciplinary constructions such as 'information seeking,' 'information encounter,' 'information professional,' 'information search,' and so on, occur instead in the paper's Introduction and Discussion, where explicit connections are made to other research studies and issues of interest for professional and applied work in the field. For example, the purpose of the Discussion is to suggest some 'implications for the information search process,' such as how information systems may be improved, the role of intermediaries in assisting novices in finding pleasure reading, and the generalizability of the study to 'goal-directed information seeking situations' (796).

The uneven distribution of Ross's uses of 'information' suggest an answer to the question that remains upon reading only the descriptions of her cases: What makes them examples of *information* encounters? Disciplinary imperatives aside, the only reason to invoke the term is that the effects her subjects experience are produced by encounters with *documents*. If it were not for the *books*, there would be no reason to apply the term, just as it would not be employed when speaking of the transformative effects of religious experience (if we are transformed by reading holy documents, we do not call these 'information encounters'). We are used to thinking of information as the content of documents, so in these cases, when powerful effects are realized in reading books, even just for pleasure, the use of the term 'information' can sometimes get a grip, however tenous. The weakness of Ross's commitments to realism about information can be measured not only by the lack of explicit concern for what information itself might be, but also by the reliance upon no component of the realist idea of information other than the concept of being conveyed by documents. The strength of commitments to nominalism about information can be measured by Ross's extension of 'information' to cases few outside and likely even inside the discipline of library and information science would recognize as such.

Ross's paper illustrates strategic uses of 'information' to situate research studies within a widely diffused body of work. Because the highly interesting stories she recounts have much to say to studies of reading, even if the word 'information' were never used, their research significance to information studies has to be constructed. Her uses of 'information' show how this can be done: the word is almost always deployed to forge discursive links to other studies in sections of her paper that do not describe her cases but embed them in theoretical interpretations. Such documentary practices are common to the discipline. Studies of information systems, information retrieval, information needs, information uses, information seeking, information encounters, and information behaviour are brought together under the umbrella of 'information' not because of real similarities between the cases studied and supported by a theory of what information itself might be, but because, in typically nominalist fashion, there is a disciplinary and scholarly practice of using the term to interpret studies of issues as disparate as searches for bibliographical references in large databases and library catalogues, use patterns revealed by World Wide Web transaction logs, how social science faculty researching stateless nations describe their strategies for finding research resources, how students looking for sources to write term papers feel at various stages of the process, how women pregnant with twins gain knowledge relevant to their situation, how pleasure reading makes a difference in the lives of readers, the nature of conversations about human services between seniors and nurses at foot clinics, and many, many more. Ross's uses of 'information' show how to apply a well-accepted and widely used disciplinary *form of description* as a theoretical marker that situates one's work in the diffuse and widely dispersed field of information studies. The publication in *Information Processing and Management*, a journal renowned for its highly technical papers in information science, of a study of reading for pleasure interpreted as information encountering testifies both to the success of Ross's project and to the strength of disciplinary and professional tolerance of nominalism about information.

The closer studies of information behaviour move towards explicit concern with theoretical issues, particularly those seeking to find some common ground for a dispersed field, the more realist tendencies about information emerge, especially in epistemic and cognitivist guises. The work of T.D. Wilson, the originator of the idea of information behaviour, is perhaps the best example. Two of his theoretical concerns make the case. The first is his plea for shifting the theoretical focus of information-seeking studies from information to communication (1994), and the second his survey of models in information behaviour research (1999).

Wilson has always been clear about his commitment to the cognitive viewpoint in LIS.[16] It both underlies his proposal to shift the discipline's primary theoretical object to communication and demonstrates how little is involved in such a shift when mental activities are central to the analysis. His 'integrative model' aims to impose some theoretical unity on the chaos of fifty years of information needs and uses studies. Its underlying conception, whether framed in terms of information or, as he would prefer, in terms of communication, is based upon his phenomenological interpretation of Dervin's sense-making model. The essence of information seeking, information use, and information need is located in a conception of an individual whose passage through time and space is guided by the mental construction of her 'own social "world" from the world of appearances' (1994, 32). This humanistic reliance on the unity of the individual subject is even more pronounced in Wilson's commitment to a 'drive to seek information' – a 'basic need in the individual' (34). Thus a mental process of constructing a private social world from appearances, driven by a basic need, is postulated as the underlying theoretical object unifying various 'needs' and 'seeking' behaviours. Information behaviour is driven by a search for sense. Minds construct sense by building internal social worlds from appearances. The problem of how information is communicated addresses how useful epistemic content gets into the information-seeking mind. Like other conceptually biased models, it construes the communication of information as traffic in mental representations. Communication, Wilson's proposed theoretical kind, incorporates the essentials of the idea of information as presence of content to a mind. His interest in a theoretical unification of disparate studies in information seeking lead him to a robustly realist and cognitivist conception of the nature of information.

We see the same interest at work in his survey and comparison of models of information behaviour, whose aim is 'to discover how they may relate to one another and, perhaps, propose an integration of the models into a more general framework' (1999, 249). A realist solution to this problem is to be expected from the close ties between seeking and communicating information and information behaviour, which Wilson defines as 'those activities a person may engage in when identifying his or her own needs for information, searching for such information in any way, and using or transferring that information' (249). His realism is also announced in his interpretation of a 'model,' as evidenced in his concluding questions for further research: 'to what extent are the different models complete, or reasonably complete representations of the reality they seek to model?' (267). Wilson's own 'problem solving model of the information seeking and searching process' is intended 'to provide a kind of linking or integration of at least some of the models discussed' (265). However tentatively offered, it seeks to represent an internal reality: the sequence of mental events – the goal-directed transitions in mental states from uncertainty to its resolution – occurring in the minds of individuals. Information is therefore a singular kind of thing: that which individuals seek in order to resolve uncertainty.

Wilson's motivation for seeking a singular kind of thing as the object of all information seeking is understandable. An analogy helps. I suspect there are many who might think that seeking deer would be a process well worth investigating, and one could even draw diagrams of 'deer-seeking behaviour' – 'models' of such behaviour. One could do the same for seeking gold in streams, but there would be little point in comparing such diagrams or 'models,' as Wilson does with respect to 'models of information behaviour,' because seeking deer and seeking gold are searches for different kinds of thing: there would be no reason to suppose that any common features of the seeking reflected shared properties of any theoretical interest. If what one seeks or encounters in studies of seeking or encountering information is different in each case, there would appear to be little rationale for Wilson's exercise of superimposing and comparing diagrams purporting to 'model' the various searches, especially if models are intended to represent reality as completely as possible. Common features of models of seeking different kinds of things are of little use in providing the wanted theoretical unification a robust realism brings. They tend to be too general and therefore not very useful as bridges between the models, as Wilson rather tactfully notes in his comments about the categories of 'starting' and 'ending' in Ellis's model – categories which could be applied as easily to 'models' of deer- or gold-seeking behaviour, not to mention looking for yesterday's left-over dinner in the recesses of my refrigerator. Wilson's work on models of information behaviour is a fine exemplar not only of the close connection between a robust realism about information and yearnings for a theoretical unification of disparate studies, but also of the tenacity of mentalistic interpretations of information itself.

Wilson's anxieties about how to unify many disparate studies resonate throughout the literature on information seeking and behaviour. In the concluding remarks of his comprehensive book on research in the field, Donald O. Case comments on the problems of diversity: 'there is more *fragmentation* than ever before among the thousands of diverse examples of human information seeking' (2002, 286), a situation that leads him to muse, 'Certainly we could say that information behaviour research has become more "scholarly," but perhaps also more pointless as well' (287). After noting, in a chapter devoted exclusively to the topic (39–63), the many definitions of 'information' offered in the literature, Case is led to the conjecture that there need not be one for research to proceed, a conclusion certainly supported by the evidence and the many reminders throughout his text that one rarely finds such definitions, or definitions of related concepts such as information seeking or information behaviour. He is content to treat information as a 'primitive' concept, by which he means a 'useful conceptualisation,' or 'a phenomenon that we all recognize when we see it in its various forms' (59). But 'primitive' concepts have histories and cultural heritages, are rarely conceptually coherent, and cannot be legitimated by appeals to 'recognition.'[17] Not content to leave the idea completely unexplicated, Case tellingly relies on mentalistic conceptions: 'in this book "information" will be taken to mean *any difference that makes a difference* to a conscious, human mind – this *perceived difference* is a 'unit of mind' that must be *inferred* (40); '[f]or the purpose of this text, we will assume that a conscious brain must be engaged at *some point* for information to be said to exist' (60); evidence for the observation that 'information seeking research has at last come into its own' is that there are now many more 'attempts to "get inside the head" of the seeker' (287). And, although the last of Case's eight 'lessons of information behaviour research' is that information behaviour is not always about sense making, he does claim that Dervin's sense-making 'paradigm' is the single most significant factor in the profound change he claims to have taken place over the past quarter-century in information-seeking research. He understands sense making in mentalist and representationalist terms: it is about the behaviours individuals engage in to move between 'subjective, internal information' – which 'represents our picture, or cognitive map of reality' – and 'objective, external information,' which 'describes reality' (43). Case's 'primitive' idea of information – 'whatever appears significant to a human being, whether originating from an external environment or a (psychologically) internal world' – is a robustly realist conception, one which draws together representationalist themes reaching back to the Cartesian origins of our modern concept of mind with much more recent cognitivist and mentalistic tropes. The skill with which he puts this conception to work in unifying the staggering amount of research his book addresses is not only a tribute to its enduring presence in information studies, but also a major contribution to the construction of whatever discursive stability it may continue to enjoy.

The Beleaguered Journal Article

The adventures of epistemic content narrated in the *ARIST* reviews and information user studies, whether the study subjects are scientists or everyday life information users, seekers, or encounterers, tell tales of information intersecting with the minds and souls of a great variety of individuals, whether conveyed by documents or received directly from observation of the world. We see scientists working inside their own heads, processing the information needed to derive new results according to their own specialized method. We see information scientists navigating World Three. We see individuals making sense by seeking information to fill out internal pictures or cognitive maps of their everyday realities. And we see others actively seeking knowledge or, for better or worse, bumping into it serendipitously. We hear people giving, receiving, and sharing information in conversation, searching for it to find answers to questions, to identify problems and solutions, and to resolve uncertainty. We catch them at their pleasure reading, watch them gain courage, coming to accept others and confirming their self-worth. In these narratives, we encounter information as a theoretical kind, either explicitly, furthering the realist ideal task of harmonizing disparate studies, or implicitly, silently countering the fragmenting tendencies of nominalist uses of the word and shoring up the tentative bridges such uses erect between disparate research projects.

In the following chapters, I try to show that there are different tales, tales of documentary practices, their effects, and of the social circuits in which documents travel. These are tales without information. My route to them begins here, by returning to the paradox of scientific documentation mentioned in the Introduction. There, I suggested that the paradox, which imagines the formal literature of science as both the most and least important resource for the production of scientific knowledge, arises from allegiances to a conception of information as the epistemic content of documents. On the one hand, such a conception supports the idea that the information conveyed by the journal article contributes to the best representation of the natural world that science as a collective enterprise has to offer at any particular time. Coupled with the idea that the primary product of scientific work is the published paper, this view leads to the conclusion that science's formal literature *is* science itself. According to this line of thought, nothing could be more important to science than the journal literature. On the other hand, the idea of epistemic content has encouraged researchers in information studies to seek knowledge of the relative efficiencies of channels of scientific communication. In this arena, key questions for the journal article become: What is its contribution to the derivation of new research-front results in experimental science?[18] What kind of information does it convey most efficiently, to whom, at what stage of the research process, and in which scientific tasks? How does its efficiency in communicating information rank among the many channels of science information systems?

Questions of this sort have preoccupied researchers since the first investigations of scientific information. Empirical studies have delivered a verdict, and it is not good. At the primary site of knowledge production, the journal article conveys little if anything scientists need to produce new results. The reasons are plain: the role of any vehicle of scientific communication is to convey epistemic content, science is fast, information ages, and the system of journal literature is slow. It would appear that scientists use journal articles mainly in tasks not directly related to the production of advanced research results. Even though, as Charles Bazerman notes, 'reading consumes a substantial part of a research scientist's working life' (1988a, 235), research on scientific scholarly communication has affirmed and reaffirmed the conclusion that research scientists rely very little on the journal article in the conduct of advanced laboratory work.[19] Nor do they make much use, in this (and often any other) aspect of their work, of formal access and retrieval systems. Based on her ethnographical studies of the high-energy physics community, Sharon Traweek's observation speaks for many: 'Good experimentalists do, write, and talk physics, but they rarely read physics. Important results are usually written up quickly and are available as "preprints" within a few weeks of discovery or innovation in detector or software design; preprints appear as journal articles within months. Particle physics changes so rapidly that waiting to learn of interesting data, detector innovations, or new theoretical developments until they appear in the journals is regarded as exceedingly unwise. What is being talked about is the current, more advanced knowledge; what has been written is considered established, uncontested, and hence uninteresting' (1988, 120–1).

Schemes for replacing journal publication with speedier systems have been a feature of research on scientific documentation at least since the time of Otlet, as we saw in chapter 1, and they surface today in proposals to disseminate research findings on the Internet and the World Wide Web.[20] Garvey's research with Lin and Tomita addresses the 'usefulness of information published in journal articles' directly. It was, he writes, 'directed at a special class of information-consumers – workers active on the research front associated with the specific subject matter of the articles.' He points out that the 'results of the study clearly show that most such workers had gained information useful in their work before such information was published in journals ... These findings are not too surprising in light of data which strongly suggest that the scientific information in most journal articles was six months-one year behind the research front at the time the articles were published' (1979, 223). A similar but typical assessment was issued by Price, one of the founders of studies of modern science information systems. After noting that '[o]nly incidentally does the paper serve as a carrier of information' (1986, 61), he is so bold as to predict the demise of the journal article: 'after three centuries the role of the scientific paper has drastically changed. In many ways the modern ease of transportation and the affluence of the elite scientist have replaced what used to be effected by the publication of papers. We now tend to communicate person to person instead of paper to paper. In the most active areas we diffuse knowledge through collaboration ... All this makes for considerable change ... It has made the scientific paper, in many ways, an art that is dead or dying' (80).

Just because journals fail to deliver the information needed at the research front does not mean that they are irrelevant to all scientific tasks. Since scientists read a lot, there must be some kinds of thinking engaged by the epistemic content of these documents. Having investi-

gated the problem, Garvey echoes the findings of many other researchers, which point to the conclusion that the use of journal articles is an
inverse function of cognitive task level: 'Some other worthwhile functions of journals are: (a) general awareness of a field, mainly for those
persons not primarily involved in a specific subject-matter area of a
field; (b) current awareness of a field, mainly for those persons who are
not conducting research in the field and who have no other access to
the research front; and (c) certification of the knowledge of a field,
mainly for those persons who want to relate a specific piece of work
suitably to other scientific work' (1979, 223). Scientists consult the formal literature when they step outside their laboratories to inform
themselves about what they currently do not know: 'The scientist, as a
consumer, uses this body of material to acquaint himself with the accumulated knowledge that exists whenever he embarks on *new* research
or inquiry ... This material is also the basis for instruction of future
scientists' (257–8).

The journal article also has other important functions, as suggested
in chapter 1's discussion of Merton. The view that articles function as
priority claims serving to establish a social hierarchy among scientists
need not require that one abandon the idea that articles convey information. One may hold instead that they simply do not convey information of epistemic value at the research front, but of significant
epistemic value in coming to know which scientists enjoy a particular
level of social standing in scientific society. For Price, the issue is clear:
'scientific communication by way of the published paper is and always
has been a means of settling priority conflicts by claim-staking rather
than avoiding them by giving information' (62). He observes that such
devices are an important part of the scientific enterprise: 'claims to scientific property are vital to the makeup of the scientist and his institutions. For these reasons scientists have a strong urge to write papers
but only a relatively mild one to read them' (62). Garvey concurs, in
equally apodictic terms: 'There seems to be little question that scientists use journals as a means of gaining and maintaining visibility
among their scientific peers ... scientists use journals to identify pub-
licly the priority of their discoveries' (257–8).[21]

We are left, therefore, with the dilemma of the scientific journal article. Due to its age, size, and perhaps even quality, it has been shown to
be deficient in conveying the information needed in the performance
of advanced scientific work. Moreover, studies also show that the
massive systems apparatus, including human intermediaries such as

librarians or other information professionals dedicated to the organization of and access to journals, is largely ignored by active research scientists. The more one seeks information for advanced scientific work, the less one uses the scientific journal article and its organizational apparatus.[22] The utility to research science of the journal literature therefore varies inversely with the amount of effort expended in its organization, representation, and administration. The degree of use of information services, apparatuses, and procedures turns out to be a function of how little rather than how much knowledge users possess. The general picture, paradoxically, is of a massive apparatus for beginners, students, and junior scholars, as well as for tasks unconnected with advanced research but useful for initiating work in new fields, for browsing, or for 'keeping current.' The journal article communicates very little, if anything, of substance contributing to the performance of research science, perhaps communicating only a subtext about science's social systems of intellectual priority and status hierarchies. It is therefore left in an unhappy and exposed position. Epistemological interpretations of information as the epistemic content of documents leave a very important question: What is the point, purpose, and role of the scientific journal article? Putting the article to the epistemological test, we discover that research science does quite nicely without it. Not wishing to abandon our epistemological devices and the privilege they bestow upon the communication of information, we look for it elsewhere: 'information' remains a signifier for the epistemic content of science, and the circulation of this content continues to motivate the study of scientific communication.

Whether we actually find in other channels what we sought in the journal article will not be pursued in this book. My motivation for revisiting the paradox of science's formal literature here is to suggest that its documentary significance, as measured by the volume, structural complexity, and organizational apparatuses of its literature, is related in some important way to the scientific enterprise itself. The paradox shows that alternative approaches to its role in scientific practices are called for. The argument developed in the following chapter aims at attaining a happier analysis of its role by abandoning the privilege granted to epistemological concepts of information, communication, and knowledge in studies of science's documentary systems. I argue for deflating the idea of information, not only as epistemic content, but also as a theoretical kind, proposing a turn to scientific practices, which intervene in the world rather than represent it.[23]

3
Epistemology versus Practice

The previous chapters have shown that from the time of Francis Bacon to the present day schemes to increase scientific productivity have centred upon proposals to reform and even overhaul science information systems. If only we could discover, in the complex information environment of scientists, the really crucial bits of information which are absolutely necessary to solve their research problems, we would then be able to identify their sources and optimize the use of the channels conveying them. Bacon's *New Atlantis* and Otlet's Universal Book make their ideal systems dependent upon rigorous techniques of document production and organization. Scientists' behaviour is disciplined such that their results are documented in the form required to maximize the efficiency of knowledge production. Today, some think that given the potential of new information technologies, a virtual *New Atlantis* is a real possibility.[1]

Since an important aim of this book is to challenge the central role of information and its communication in scientific activities, I would like to insist on the question, why is so much attention paid to designing, refining, improving, building, and rebuilding science information systems? I do not mean to challenge the need to produce and improve specific tools, such as handbooks, reference works, or databases of numerical, visual, and textual data. Procedures for developing such tools are well developed, and users find them more or less suited to their activities. The question is directed instead at the idea that science and information share a fundamental and essential connection, and that doing science is in some intrinsic way an engagement with information. It is also directed at assumptions that the use and communication of scientific information must in essence be systematic activities, and that unearthing their underlying principles permits the often recalcitrant and messy world of information exchange to be optimized.[2] For it is hardly self-evident that the totality of the science information system needs to be improved. Nor is it obvious that such a totality even exists. There is something metaphysical, even strange, about the idea that scientific knowledge production depends upon maximizing efficiencies of the communication of information. The oft-expressed belief that science's information system is at least in a state of disrepair, or worse, at risk of catastrophic collapse, is rarely supported by evidence or example. It rests instead upon the conviction that at least some if not a great quantity of epistemically valuable information *must* be lost in 'the mess of unreadable scientific literature,' as Bernal put it. But convictions modalized as necessities are symptoms of the metaphysical imagination, not reasons for beliefs.

All the attention paid to improving information systems seems somewhat odd when knowledge production is seen less as a matter of producing, processing, and exchanging information than as making things work in the laboratory and manipulating material things, processes, and techniques. Scientists would appear to have been quite successful at such work since the sixteenth century, and even before, but especially today, when their labours enjoy huge military and corporate support. If the communication of information really plays a central role, would the success of science not suggest that perhaps scientists have been equally successful at developing and using the information systems they need? What then, is the source of the conviction that a perfected science information system would *necessarily* increase scientific productivity? How are science, scientific knowledge, and scientific information conceived to make this a plausible idea?

This chapter suggests that information becomes central to the production of scientific knowledge when science itself is interpreted epistemologically. The previous chapter showed that studies of scientific information use, and studies of information use generally, exhibit to varying degrees a conceptual and cognitivist bias in which information is understood in terms of concepts such as the epistemic content of documents, representation, presence to consciousness, immateriality, and disembodiment. This abstract idea of information assumes a privileged status in proposals for improving scientific work when the nature of such work is understood in terms of the same concepts and their cognates. As presence to the scientific consciousness, 'information' is just another term for what the scientist produces, processes, and communicates. Consequently, the urgency of improving science information systems gains support from epistemological descriptions of scientific activity. Moreover, because epistemological discourses of science privilege the systematic structure of scientific knowledge, they support the idea of a *system* of scientific information, providing a legitimating discourse for science information systems as objects of theoretical knowledge and disciplinary inquiry. This discourse helps establish in the scholarly and disciplinary imagination of information science an image of information system design in aid of efficient flows of information to sites where knowledge is generated.

Epistemological descriptions of science also provide discursive resources for articulating information as a theoretical kind – a singular kind of thing, process, or phenomenon about which theoretical knowledge may be gained. In what follows I call an *information and communication (IC) model* any model of an activity or phenomenon which privileges information and its communication by virtue of treating information as a substantive denoted by the term 'information.' Thus the distinguishing feature of any IC model is that it deploys information and communication as its primary, substantive, analytical, and explanatory resources for representing some phenomenon, thing, or activity.

Information and communication models explain not only science and scientific activities. Education and labour in the 'information age,' the formation of cultural identities in cyberspace, leisure activities, personal relationships, capital accumulation, and even warfare can and are described by IC models. A common 'pop psych' view of human relations, including sexual relations, interprets them in terms of the communication of information. Having trouble with your partner? Perhaps it's the message you're sending. Having trouble finding a partner? Perhaps your wardrobe communicates the wrong information. When bombs fall on Iraq, generals speak of 'sending a message.' An epistemological discourse of science is just one example of an IC model. My argument here may be stated as the claim that an epistemological discourse of science is an IC model of scientific labour according to which the activities of producing, processing, and communicating information are central to science itself.

The point of investigating these connections is to show that they are not intrinsic, essential, or necessary. If science may with justification be seen not primarily as a conceptual practice, and the product of scientific practice not as a conceptual field, information loses its privileged status. And when the ideas of representation, epistemic content, and even theoretical kinds are abandoned, the idea of information itself as a disembodied, immaterial entity conveyed by the channels of an information system is deflated. The role in scientific activities of documents, such as the journal article, will then have to be sought beyond their capacity to communicate the information IC models assume to be the life-blood of research science.

The next section of this chapter contrasts the epistemological discourse of science with an alternative: a discourse of practice.[3] This alternative articulates the material aspects of scientific activities. It reveals the localized, contingent, and difficult laboratory labour involved in producing stable laboratory phenomena and extending them to other research contexts. In contrast to the unified conceptual field of knowledge featured in epistemological discourses of science, discourses of practice reveal the disunity of the natural sciences. Scientific activities are dispersed among many and diverse fields of scientific labour. A focus on practice emphasizes the work involved in aligning many heterogeneous elements of scientific culture to stabilize scientific phenomena. A discourse of practice also challenges IC models of science. When scientific activities are described in terms of practice, labour, and materiality, neither information nor communication appear as theoretical kinds. Instead, these concepts are deflated by a nominalist interpretation according to which 'information' and 'communication' each refer to an open-ended series of instances, with no unifying theoretical connotation.

My aim is to show how the shift from epistemology to practice transforms epistemological questions about the communication of information into material questions about the production of statements and the role of occasioned utterances, inscriptions, or graphical images – that is, documents – in the diverse practices of scientific culture. Questions about the role of a specific documentary form, such as the journal article, therefore become questions about the ways in which occasioned inscriptions with specific institutional properties are aligned with other elements of scientific culture to produce laboratory phenomena with sufficient stability to become scientific facts, or specimens of scientific knowledge.

Scientific Knowledge: Conceptual Field or Construction Zone?

Epistemology

The title of Garvey's book, *Communication: The Essence of Science* (1979), could serve as a banner for the epistemological approach to the significance of any particular document form in scientific activities. If communication is the essence of science, then whether a particular document form is essential to science itself will depend upon the degree to which it is used to convey scientific information. Channel use studies are seen as more or less straightforward empirical inquiries designed to settle the issue. Thus Garvey's dictum provides a framework for investigating the role of the journal article in scientific practices. If empirical work on channel use shows that the journal article is not used to convey scientific information, it follows that it is not essential to science itself.

Chapter 2 has shown that empirical studies of journal article use conclude that it fails to convey the scientific information wanted at the research front. Thus the question of whether communication by journal article belongs to the 'essence of science' requires that the information it conveys be partitioned into that pertaining to essential scientific activities and that pertaining to non-essential ones. Science is a huge enterprise; many channels of communication are open, and many kinds of information are transmitted. Which kind is essential to science? Does background reading count? Does brushing up on new topics or reading in preparation for delivering a new graduate course count? How about the information that conveys the priority claims that position authors in the hierarchy of science's social structure? The reactions to the failure of the journal article to deliver research-front information canvassed in the previous chapter, and warnings of the crisis of the formal literature expressed by critics such as Bernal and Bush, suggest a unanimity about such questions. The kind of information seen as most important to science itself is that which is used to produce new knowledge. The criteria used to partition what is conveyed by any vehicle of scientific communication into essential and non-essential information are epistemological. According to epistemological conceptions of scientific activities, the production of new knowledge at the research front – the primary site of scientific knowledge production – is essential to science because theory is the primary product of scientific work. To researchers and critics, the journal article is in crisis precisely because whatever it conveys does *not* belong to the essence of science. Their verdict, which relegates the formal literature to the margins of scientific work, demonstrates the strength of epistemological discourses of science.

The claim that science yields knowledge is unobjectionable, indeed, almost tautological. The distinguishing feature of an epistemological discourse of science is not that it makes this claim, but that it interprets the processes of coming to know in cognitivist, mentalist, and representationalist terms. Commenting on the taken-for-granted link between knowledge and representation, Andrew Pickering writes, 'within the traditionally restricted vision of science-as-knowledge, the representational idiom is more or less obligatory – what else can one ask of knowledge other than whether it corresponds to its object?' (1995, 5). The 'representational idiom' is an epistemological discourse. It speaks of knowledge as something like a picture, whose truth consists in the accuracy or adequacy of its depiction of its object. On this view, scientific knowledge becomes 'a mirror of nature' (Rorty 1979) or a faithful representation of the natural world. Because it is meant to be universal, it is a highly abstract representation, which consists in networks of theoretical propositions. 'Science itself,' or 'the essence of science,' is understood as a conceptual field, in the form of a logically unified set of theoretical propositions which represent the fundamental principles governing the behaviour of natural phenomena. The representations making up the conceptual field of natural science are not, therefore, material entities. They are not found among the objects, living beings, materials, equipment, instruments, and apparatus that scientists routinely manipulate. Instead, they inhabit scientific thought. Epistemological forms of description of scientific activities, common to the philosophy of science before the impact of the social studies of science, speak of propositions, concepts, hypotheses, and the relationship of models to the world. Scientific activity is described in terms of mental acts and cognitive processes: observing, conceptualizing, verifying, reasoning, inferring, deducing, hypothesizing, conjecturing, and refuting. An epistemological discourse of science is an idiom in which science is presented in terms of knowledge, theoretical representations, and abstract, immaterial objects of thought and thought processes.

The previous chapters have shown that there is a discourse in which information is described in similar terms. The bridge between the two discourses is the idea of epistemic content, perhaps best expressed in Bernal's identification of what scientists acquire through experiment or thought alone with what they acquire through documents. Epistemological discourse describes what is present to the minds of scientists, whether they observe phenomena of nature directly or by experiment, or contemplate the theoretical conclusions derived from them, in the same terms as we have seen the epistemic content of documents described in previous chapters. Bernal's claim that what is present to the mind is the same in all cases illustrates the main point of this section: that the abstract idea of information is among the discursive resources swept up in the wake of epistemological discourses of science. The epistemic content of documents – their 'information' – is indistinguishable in respect of its presence to consciousness from what an epistemological discourse of science imagines as objects of scientific thought. Information belongs among these objects: it is one of the raw materials of science's theoretical representation of the natural world. It belongs to the same theoretical kind as objects of scientific thought described by epistemological discourses.

Since science advances collectively, its information must be communicated. Assessments of the relationship between observations and theory require many observations, and a single scientist is not in a position to produce all of them. Moreover, Bernal noted that scientists now rely more on the content communicated by documents than on observation of nature. Popper's (1965; 1968; 1972) system of 'conjecture and refutation' is one model of science's collective enterprise. Epistemologically interpreted, communication becomes traffic in 'information.' When the important thing about documents is their epistemic content, they and their system of communication belong to the same epistemological interpretation of science. Serious constraints are therefore placed upon the importance of documentary practices to scientific activities. The delivery of their epistemic content is the only contribution documentary practices can make to knowledge production. Historical questions arise. Has the role of a particular documentary practice changed over time? Garvey is led to ask just such a question of the journal article. He notes that the conclusions of his studies with Lin and Tomita 'raise some questions about the function of current journal articles: Can the journal article any longer be regarded as a vehicle which effectively communicates current scientific information? If not, can the journal article be reworked to function more efficiently in the capacity of integrating scientific information into a larger framework?' (1979, 223–4). These are the same questions Bernal posed forty years earlier. Such questions belong to epistemological talk about science.

Practice

In the introduction to his important anthology *Science As Practice and Culture*, Pickering describes a study of science that privileges epistemology: 'For the logical empiricist, say, scientific culture consists in a field of knowledge and knowledge claims, and scientific practice consists in the appraisal of conceptual knowledge claims against observational knowledge, an appraisal ideally governed by some logic or method' (1992, 3, note 2). His book collects papers that present an alternative way of studying science, because what he calls the traditional science-as-knowledge approach is not helpful in investigating what scientists actually do. He asks whether 'analytic repertoires developed in the service of a problematic of knowledge can serve as the primary basis for understandings of practice,' and concludes that 'most scholars who have taken it as their task to get to grips with scientific practice in some detail have found that they cannot' (6). Even the early attempts to move beyond activities such as generating and appraising knowledge claims by coming to grips with actual scientific practices were nonetheless dominated by the predominantly philosophical science-as-knowledge approach. The shift in the 1970s to studies of the micropractices of science, or 'the move toward studying scientific practice, what scientists actually do, and the associated move toward studying scientific culture, meaning the field of resources that practice operates in and on' (2), was handicapped by its narrow focus on the problem of modelling 'the technical culture of science as a single conceptual network' (4). But as soon became apparent, this model 'does not offer much purchase upon the complexities evident in the nearest laboratory' (5). Even as late as 1985, Michael Lynch could begin his ethnomethodological study of laboratory practices with the observation, 'it is curious that the place where the *labour* of science is performed has rarely been visited in studies of science and investigated in terms of its productive relation to the material domain of scientific inquiry' (4).

The essays in Pickering's anthology represent a deeper engagement with scientific practices, and a clearer recognition of the need to abandon epistemological discourses. His book is part of a large and growing literature which emphasizes labour, the many complexities of scientific work, both within and beyond the laboratory. It signals a turn to studies of science as *culture*: 'Whereas one could once get away with thinking of scientific culture as simply a field of knowledge, in what follows I take "culture" in a broad sense, to denote the "made things" of science, in which I include skills and social relations, machines and instruments, as well as scientific facts and theories. And then I can state that my abiding concern is with scientific *practice*, understood as *the work of cultural extension*' (3). Studies of science as practice and culture emphasize the heterogeneity of the kinds of cultural elements

Pickering mentions. When contrasted to the homogeneous cognitive activities featured in the 'field of knowledge' approach, the shift to practice brings to view disparate kinds of institutions, labour processes, and material practices. The range of scientific activities is expanded far beyond cognitive processes such as generating, appraising, and forging logical relations between propositions or knowledge claims by the exercise of scientific reason. A shift of focus from cognitive to labour processes reveals scientific work as the construction of localized assemblages of things, persons, devices, and social relations.[4] These include discursive objects, such as the many symbolic representations found both inside the laboratory and in the scientific literature.

The move away from the language of 'thought and theory' has also called into question the assumed transparency of 'the communication of scientific information.' Thick descriptions of laboratory practices fail to locate science's theoretical representation of nature that figures so prominently in science-as-knowledge models. Missing too are the abstract, immaterial, conceptual, and cognitive elements and processes so central to epistemological discourses, among them the gathering, processing, and communication of information as it is understood in studies of science information systems. The close connections between epistemological conceptions of science and information imply that a shift towards practice and materiality, and away from representation and epistemology, brings with it a corresponding shift away from information and communication as significant concepts for understanding scientific practices.

The connections between the epistemological discourse of science and a corresponding discourse of information have been noted in studies of scientific practices. When 'scientists figure as disembodied intellects making knowledge in a field of facts and observations' (Pickering 1992, 6), we are presented with an 'algorithmical model' of knowledge, one 'very much in accord with the view of the information scientist' (Collins 1992, 75), who 'views knowledge as the sort of information that enables a computer to carry out its programmer's intentions.' When knowledge is seen as 'a set of formal instructions, or pieces of 'information'' (57), the salient activity of science becomes *information processing*. Scientific method becomes the program generating propositions from scientific information, and scientific communication becomes traffic in epistemic content. Rouse recognizes that the occlusion of scientific labour stems primarily from the idea of epistemic content: 'the abstract and immaterial content of knowledge which philosophers put forward as rationally arrived at or approximately true is removed from any assessment that considers scientific achievements as materially and socially situated' (1996, 22). By contrast, studies of the material activities of construction, labour, work, and practice lead directly to an engagement with issues of social structure, organization of labour, economy, politics, and power.[5] If representations are abandoned in favour of heterogeneous practices engaging a wide variety of things, persons, and social relations, then political issues do not remain external to a cognitive interior of science but contribute directly to the production of scientific facts and phenomena. The shift to practice signals an end to what Rouse calls the 'normative anemia' of science studies (23).

Studies of scientific practices translate the question about the role of documents in science into questions about the relationships between particular document types and the many other disparate elements of scientific culture assembled to construct stable scientific phenomena. Bazerman sees the significance of the role of discursive practices in these assemblages: 'this notion of system can be extended to include all kinds of symbolic representations, relationships, practices and objects that must be brought into alliance for any technology or scientific knowledge to take hold' (1994, 118). The inclusion of discursive practices and objects sheds light on the more general question of 'how the process of linguistic construction is intertwined with social and material practices and relationships' (119).[6] A discourse of practice replaces the question, what is the role of the scientific journal article in communicating the information scientists need in order to advance scientific knowledge? with another: What is the role of the journal article in stabilizing scientific phenomena and contributing to the maintenance and development of research programs?

The important differences between discourses of practice and epistemological discourses can be summarized as follows:

1 Epistemological discourses articulate science as knowledge and describe knowledge as an immaterial conceptual field whose primary directive is to yield a theoretical representation of the natural world; by contrast, discourses of practice articulate science materially, as a kind of construction zone.
2 Viewed epistemologically, knowledge is discovered; viewed in terms of practice, knowledge is manufactured (Knorr-Cetina 1981).
3 In discourses of practice, activities are varied, skilful, craftlike, and

local; they engage a wide variety of material, social, and discursive objects, with the aim of producing more-or-less stable assemblages of things, people, and documents, which can be extended beyond the laboratory as resources for other research contexts and transscientific fields. In epistemological discourse, typical scientific activities are universal and rational, taking place primarily in individual minds, and becoming for the most part exercises in logic, since their primary aim is to generate and establish evidentiary relationships between observational and theoretical propositions; developing and testing theories by observation gained through original experiments and their replication are taken as paradigmatic and fundamental activities of science.

4   The objects engaged by epistemological interpretations of scientific activities are immaterial: meanings, truths, propositions, representations, the epistemic content of scientific documents, and information. Discourses of practice address the strategies, deployments, appropriations, and manœvres involving discursive objects; these are articulated as material things: occasioned utterances or inscriptions, among them statements appearing in documents, conversations among colleagues, records of laboratory conclusions, readings displayed on laboratory instruments, computer-generated data analyses, and many others, whether generated by monitoring devices in the laboratory or by the historically contingent, institutional imperatives governing the production of scientific literature.[7]

Chapter 1 opened with the statement that there are close connections between ways of thinking about scientific documentation and ways of thinking about the nature of the scientific enterprise. Descriptions of the role of a particular document type in 'science itself' will depend upon which of the alternative discourses the analyst adopts. If the discursive resources used to articulate the categories of knowledge, representation, immateriality, and epistemic content are undermined by a discourse of practice, they lose their utility in articulating the importance of documents to scientific practices.

Deflating Information

When epistemological discourse employs the terms 'science,' 'knowledge,' and 'information' as grammatical substantives, it projects particular kinds of entities or phenomena as their referents. Science, knowledge, and information are described as general kinds of things with the coherence and singularity enjoyed by objects of theoretical knowledge. Although in their work scientists use journal articles, preprints, conference presentations, letters, and many other kinds of documents, the realist claim that they are all instances of *information use* is presented as a non-trivial and significant theoretical insight. Epistemological discourses describe science, information, and communication as *theoretically coherent kinds*.

The concept of theoretically coherent kinds appears in Rouse's discussion of philosophical debates about the nature of science (1996, 28). When its legitimation is at stake, science is described as a particular kind of thing about which general statements hold. Consider the controversy about scientific truth that visited 'science wars' upon popular culture: does science represent the natural world or are its results social constructions? Only when science itself is imagined as a theoretically coherent kind can this question drive such a debate. Rouse's discussion features a family of related theoretical kinds: epistemic content, scientific knowledge, truth, representation, and reality. His inclusion of epistemic content in this list recalls my claim that epistemological discourses of science sweep along in their wake the notion of the epistemic content of documents. Since information is imagined as epistemic content, it too belongs to the family of theoretically coherent kinds whose parent is the idea of science itself.

Epistemological discourses describe information as a particular sort of theoretically coherent kind: a representational kind. There are other ways to think of information as something about which rigorous knowledge may be gained. Claude Shannon and Warren Weaver's model of information is an example. Their interpretation of information is theoretically and mathematically rigorous, but it does not construe information in representational terms. Famously, and to some notoriously, its analysis of information in terms of signal-to-noise ratios avoids the idea of meaning altogether. It is therefore possible to reject epistemological discourses of information yet retain the idea of information as a theoretically coherent kind.

A discourse of practice challenges not only epistemological descriptions of information but also the very idea of information as a theoretically coherent kind. It does so by contesting the idea of science as a theoretically coherent kind. A discourse of practice provides a *deflationary* account of science and scientific knowledge (Rouse 1996, 28, 195–204). If there is no essence of science, and if scientific activities exhibit no unifying substratum, the very idea of any unified, coherent model of science, much less an IC model, is deflated. Furthermore, if science itself is not a theoretically coherent kind, then neither are any members of its family of phenomena. Deflating science means deflating representation, knowledge, conceptual fields, and epistemic content. Most significantly for my purposes, it means deflating information.

What is a deflationary account of science? Such an account holds that scientific practices are many and varied, historically contingent, and open-ended in the sense that what may in the future come to count as scientific activity is in principle unpredictable. Scientific practices are not unified by virtue of a common essence, whether that essence consists in the disciplined, rational pursuit of a theoretical representation of the world, the universal structure of rational-scientific subjectivity, a common theoretical language, or a social realm grounded in consensus about paradigms, interests, or norms. Neither nature, mind, language, nor society provide any essential underlying basis for thinking of science as a singular kind of thing because these explanatory unities are no less problematic than 'science itself.' Rather than reducing the essence of science to them, a deflationary account rejects the essentialist, reductionist project altogether. It therefore provides no purchase for the theoretical imagination. Rouse writes that the 'cultural studies of science reject the existence of an essence of science or a single essential aim to which all genuinely scientific work must aspire' (242-3). At another place, he puts it this way: 'there is much scientific knowledge but no nature of scientific knowledge' (180). Deflationary accounts disperse science across a wide field of historically contingent cultural practices.

When science is deflated, the term 'science' becomes merely a handy verbal device for gesturing towards an open-ended set of contingent and localized activities whose warrant as scientific derives from various and in principle always contestable procedures of justification. The device is handy even without the comfort of knowing in advance how to settle what will count as being 'like' the activities currently called 'scientific.' The assemblages of people, material things, social relations, and documentary practices that produce scientific knowledge (Rouse calls them 'epistemic alignments'; they are discussed in more detail in chapter 4) take many different and inherently unpredictable forms. Their justifications as knowledge are enacted anew in changed circumstances: 'There is no reason to believe that [historically] striking shifts in the distribution and justification of knowledge have come to an end and thus no reason to believe that the domain of knowledge (or its historically constituted subdomain of "scientific knowledge") can ever be clearly demarcated as the object of a theoretically coherent inquiry' (203).

Deflationary accounts of knowledge undermine epistemological discourses in several ways. First, both individual and communal knowing subjects are no longer privileged sources of scientific knowledge, because knowledge is interpreted in terms of practical engagements with the world, not as something possessed by knowers. Pickering's 'posthumanist' account of scientific practice describes knowledge as 'the constitutive intertwining and reciprocal interdefinition of human and material agency' (1995, 25–6). It is located not in minds but in fields of practical activity in the midst of which the knower is immersed: 'attributions of knowledge are thus more like a characterisation of a situation knowers find themselves within rather than a description of something they acquire, possess, perform, or exchange' (Rouse 1996, 133). Second, if knowledge is not a theoretical kind, it cannot be explicated in representational terms. Statements are deprived not only of the content assumed by IC models to provide raw material for theoretical representations of the natural world, but of any content at all. Third, deflationary accounts of knowledge dissolve the bond of shared beliefs or cognitive commitments that epistemological accounts assume bind together members of scientific communities. Finally, to reject the general category of representation is also to reject the notion of epistemic content and its centrality to scientific activities.[8] When epistemic content is deflated, so is the epistemological idea of scientific communication as traffic in it. As theoretically coherent kinds, information and communication have no work left to do in modelling scientific activities dispersed across an extended cultural field. In deflationary accounts of knowledge, IC models of science are idle.

Once dispersed into cultural fields of diverse practices, scientific activities stand or fall according to their own varied procedures and practices of justification. Bereft of the support of purported theoretical kinds such as epistemic content, representation, knowledge, and science itself, information and communication are likewise fragmented, differentiated, dispersed, and scattered over cultural fields of documentary practices. Studies of scientific practices show that we learn more about science if we relinquish its interpretation as a theoretical kind. We also learn more about the role of documents in scientific labour if we relinquish epistemological commitments to their informational and communicative properties, which gloss over complexities of socially and historically situated documentary practices, topics of special urgency in an era of knowledge production configured by political and ideological strategies of Big Science conducted primarily in the service of military and consumer interests. Understanding the many and varied documentary practices of scientific work does not require theoretical accounts of information and communication any more than understanding scientific practices requires the idea of unified science. As Rouse puts it, 'the historically situated and contested development of the practices themselves suffice for us to understand them' (1996, 200). Attention might therefore be usefully turned to the variety of practices in which the discursive objects collected under the rubric 'scientific information' are engaged and to the documentary circuits that trace such objects' trajectories across cultural fields. The role in scientific activities of a documentary form such as the journal article is more likely to be revealed by attending to effects of documentary practices than by labouring under epistemological presuppositions.

The Grammar of 'Information'

In epistemological discourse, IC models of science are in play. Information and communication become robust theoretical kinds, at par with science itself. But in many studies of scientific communication, 'information' and 'communication' are employed as grammatical substantives without any explicit theoretical rationale. Case (2002) has shown that atheoretical uses of 'information' abound in the thousands of information-seeking and information behaviour studies he surveyed, yet they nonetheless feature substantive uses of the term. To what extent are science-as-knowledge models, or epistemological construals of scientific activities, implicated simply by employing 'information and 'communication' as substantives? To what extent is a specific theoretical rationale for information and communication implied, assumed, or otherwise implicated merely by virtue of using these terms as referring expressions? Wittgenstein says that 'one of the great sources of philosophical bewilderment' is that 'a substantive makes us look for a thing that corresponds to it' (1969, 2). In what way and to what degree is the thing corresponding to 'information' invoked in any study of science information systems in which the term is presented as if it refers to a significant theoretical object? These questions are important because they address the strength of the connection between merely substantive uses of 'information' – the primary analytical resource of studies of science information systems – and an epistemological discourse of science. If the latter are implicated simply by the use of 'information' as a grammatical substantive, the deflationary force of discourses of practice extend beyond studies of scientific information in which such models are explicit to any which use the terms as referring expressions.

Chapter 2 has shown that information studies vary in the degree to which their epistemological assumptions are made explicit. In some, such assumptions can be uncovered even though it may not be easy to see the relative force of epistemological versus professional and scholarly discourses in constructing information as a theoretical kind. In some cases disciplinary commitments to IC models are clear, especially when information is seized upon to legitimate scholarly pursuits and professional expertise. In chapter 2 we have seen that in others, epistemological conceptions of science are in play through rationales for treating information as a theoretically significant object of scholarly concern imported from references to textbooks and other samples of science's tertiary literature. The issues raised in this section pertain to the many studies of science information systems in which 'information' is used as a grammatical substantive only, that is, those in which no theoretical rationale for information as a theoretically coherent kind is found. Is there any connection between studies such as these and science-as-knowledge models? If there is, the shift from epistemology to practice threatens them as much as it does those in which epistemological assumptions are made explicit.

We have seen that information studies investigate uses of various channels. A typical sort of case might be a study of how scientists perform subject searches in large online databases for bibliographical records of journal articles. Suppose that epistemological presuppositions or any other kind of theoretical rationales for information are absent. Suppose too, as is often the case, that there are no explicit accounts of what scientists do with the bibliographical records they retrieve. Assume that for all we know from the study, the scientists may be engaging in aesthetic pursuits, or perhaps trying to locate specimens of bad science, or to identify the high-status authors of particular scientific fields. Even though the study's topic is highly specific – regularities in the subject search behaviours of scientists using online bibliographical databases to locate records of journal articles – let us suppose, as is almost always the case and as we have seen in Pettigrew's work, that conclusions are not framed at this level of specificity, but instead, that the scientists' actions are described as instances of 'information'-seeking behaviour. They are not, we assume, described as instances of 'bibliographic-references-of-journal-article-in-large-online databases'-seeking behaviour. The use of 'information' as a substantive in drawing conclusions generalizes from specifics to invoke seeking for an abstract kind of thing: information. Even when the practical significance of the study is explicitly avowed – to improve subject search engines of bibliographical databases of journal articles – the unexplicated use of 'information' as a substantive functions as a discursive placeholder for the study's theoretical significance.

This implicit assumption becomes clear in the following example, borrowed from Karin Knorr-Cetina's (1981) discussion of a scientific article on the application of ferric chloride. To say that scientists found information, or information was communicated to them, in an article reporting the application of ferric chloride is not *by itself* to say anything of theoretical import. The statement, 'Scientists found the information that ferric chloride is a low-temperature protein-precipitating agent,' says no more than 'Scientists found that ferric chloride is a low-temperature protein-precipitating agent.' But if information is eliminible in accounts of what scientists seek, need, or use, then it plays no role, and has no significance, whether as a theoretical kind or anything else. For it to play the role of a theoretical object, there must be some difference between saying that scientists seek the solution to a particular problem and that they seek the information to solve it. If there is no such difference, this use of 'information' has no theoretical significance. The term idles: it becomes an empty signifier. Yet uses of 'information' as a grammatical substantive in information studies are meant to imply that there is an important difference, since without the term the studies deploying it sever their purported connections to other work.

There are ordinary uses of 'information' in which it is not eliminable, but they do not support its theoretical interpretation. In its ordinary, atheoretical use, to say that scientists found the information that ferric chloride is a low-temperature protein-precipitating agent is to say no more than that they learned of this property of ferric chloride from books, journals, preprints, conversations with colleagues, and so on – LIS's familiar 'channels of communication.' Such a use marks a distinction between finding it out by consulting documents and working it out for oneself.[9] There is nothing theoretical about this ordinary use of 'information,' or about the difference between finding a solution to a problem and finding information to solve it. Here, 'information' is used merely as a handy generalizing device to mark a difference between mediated and unmediated solutions to problems. This is not a realist use of 'information,' as it would be if it referred to a theoretically coherent kind. There is therefore no reason to suppose that when solutions to problems are mediated by information in this atheoretical sense the various denotations of 'information' share sufficient similarities to breathe theoretical life into studies that use the term.

In its ordinary uses the denotation of 'information' in any particular study is generally settled by context, often amounting simply to the particular channel involved; for example, encountering 'information' by browsing in a bookstore means that the study's subject stumbled across a useful book, periodical, newspaper, or posted announcement. The statement that someone finds the information that p expresses nothing more about the nature of information *in general* than the statement that someone finds that p – a statement in which the concept of information is absent. Yet its repeated uses as a substantive in information studies amounts to literary performances of research significance. The term's grammatical form suggest a referent, but a shadowy, ghostly one which is rarely brought into the light. The term can perform the work of a placeholder for continually deferred theoretical significance precisely because its meaning is not addressed.

A simple example shows how the mechanism works. When my university library promotes its services with the slogan 'Information Is Our Business,' it does not make an isolated statement. Statements do not exist in isolation. They are embedded in larger discursive formations. Many of the links between different statements are established by uses of the same grammatical substantives. My library's slogan draws upon sources of significance unavailable in earlier historical periods. Today, the use of 'information' occurs in a broad context of statements in which information is treated not only as an object of theoretical knowledge, but one sufficiently momentous to define an entire historical epoch. Contemporary readers of the slogan easily recognize echoes of the 'Information Age' and various other continually proliferating and mutating discursive contexts in which 'information' plays a starring role. Whatever force the slogan may enjoy depends upon the connections forged between it and other discourses of information vigorously circulating in politics, economics, and popular culture, especially those that its authors hope inspire interest, awe, and desire. Years ago, such a slogan would be an odd, perhaps even an eccentric way of referring merely to library holdings.

Statements employing 'information' but with little theoretical rationale thus have the place for their significance 'filled in' by connections
to sets of other statements. When the statement at issue is not a slogan
but one inscribed in a scholarly journal article, its significance benefits
from two factors: the emptiness of the signifier 'information,' and a
scholarly discursive environment already hospitable to theoretically
coherent kinds. Academic literature abounds in theoretical kinds. They
already circulate through many institutionalized channels of scholarly
disciplines. Merely grammatically substantive uses of 'information'
easily evoke theoretical significance. And when the topic is science,
prevailing representational and mentalistic cultural biases provide
strong discursive support for the idea that, for example, in using an
online database to search for bibliographical records, the scientist seeks
information, *in a theoretically interesting sense*. After all, what else could
one mean by saying that in their research work scientists seek information other than that they hunt for statements whose epistemic content
contribute to theoretical representations of the natural world?

The use of 'information' in studies of scientific systems as a mere
grammatical substantive therefore draws upon discursive resources of
epistemological interpretations of science. These include conceptions
of information as a theoretical kind. To frame research conclusions in
terms of seeking information rather than, for example, seeking references to journal articles in large, online bibliographical databases, suggests not only that the results apply generally rather than being mere
artifacts of specific channels, document types, or system properties,
but also that improving subject search engines is theoretically
grounded in general knowledge about the characteristics of a process
which is imagined as gathering raw material for the production of theoretical representations of the natural world. Thus in the scholarly narratives of scientific information the merely grammatical properties of
'information' function as links to epistemological discourses of science,
thereby transforming strategically placed uses of the term into realist
performances. Information and communication models deploy such
properties to connect different discursive fields. Grammar creates a
space for the significance of the presumed referent of 'information' to
be 'filled in' *epistemologically*. Since there flourish on many discursive
terrains powerful discourses of information as a coherent theoretical
object, merely to construct an IC model of an activity or phenomenon
benefits from conduits of legitimation between such discourses and the
phenomenon modelled. The question about the relationship between

epistemological discourses of science and IC models is not, how can they be connected but, how could they fail to be?

When Garvey argues that 'communication is the essence of science' or describes the system of 'communication of scientific information,' when Wilson calls for 'an integrative model of information need, information-seeking behaviour and information use' and suggests that 'our use of the word "information" hides the fact that the subject is actually "communication"' (1994, 42), IC models are in play. Literary performances of information and communication as theoretical kinds privilege abstract and general approaches to documentary forms. Journal articles and other document types are then easily interpreted in terms of their contributions to scientific knowledge representationally understood. But such approaches abstract from dispersed, localized, situated, and historically contingent scientific practices in which documentary forms are implicated.

The importance of documents in LIS is grounded in the imperatives of managing a range of institutionally circumscribed document types. Library management procedures are, to echo Michel Foucault, a plane of emergence of document types as objects of disciplinary knowledge. For librarianship and library science, forms matter. Their importance is retained in IC models of scientific communication, because document types convey epistemic content. When IC models are abandoned, the epistemological grounds for attention to forms are undermined. If document types do turn out to be significant (and I will argue later that the journal article – the document form of science's formal literature – is central to scientific practices), their importance derives not from happy accommodations of the imperatives of librarianship and IC models, but from studies of scientific practices.

Like system-centred approaches, person-centred information studies enact literary performances of information as a theoretical kind by strategic uses of 'information' as an unexplicated grammatical substantive. When information is deflated, there is no reason to suppose that the myriad varieties of 'information seeking' are unified by any theoretical substratum. There is no reason to suppose that 'information need' has a greater theoretical import than 'photo need,' or even 'photo-of-Jennifer-Lopez-need.' Likewise, if information is not a theoretical kind, there is no a priori reason to suppose that the results of studies of information seeking in specific media, such as electronic databases or the World Wide Web, reveal truths about a general process called 'information seeking.' Once deflated, 'information seeking' is simply a handy, ordinary, garden-variety expression gesturing towards an open-ended series of examples, *whose extension is not settled in advance*, of requests at library reference desks, telephone calls to colleagues for financial advice, dazed and confused meanderings on the World Wide Web, browsing through advertisements until struck by a 'vegetable need' for tonight's dinner, and a million others. Freed of IC models and the yearnings for theoretical significance they support, attention can turn to the many ways in which documents intersect with persons, groups, material things, and social relations, and the ways each configure the other in the process. There is no a priori reason to suppose that general concepts of informing or communicating enjoy any special privilege in understanding such interactions.

The next two chapters present in greater detail some alternatives to epistemological discourses of science. They extract implications of recent studies of scientific practices for a non-epistemological understanding of documentary practices. They show that meaning and epistemic content are not constructed by individuals, whether alone or in conversation, nor are they the kind of thing that could be made from mere representations or objects of consciousness. Like scientific knowledge, meaning is a contingent, deferred, partial, and situated *effect* of open-ended networks (Callon, Latour), epistemic alignments (Rouse), or 'mangles' (Pickering), in which persons, groups, material things, social relations, and documents are mutually shaped in local, contingent, and temporary configurations. Wilson's humanistic focus on representing subjects as sources of knowledge distances information studies from investigations of documentary practices. According to Latour, the situation is even worse, for we can learn nothing about practices by clinging to ideas of mental representations: 'the concrete work of making abstractions is fully studiable; however, if it becomes some mysterious feature going on in the mind then forget it, no one will ever have access to it' (1987, 241).

Chapters 4 and 5 reveal the limitations of epistemological discourses in articulating the role of documents in scientific practices. The central questions addressed are: Can IC models, constrained by epistemological discourses of science and scientific activities, generate an adequate analysis of the role of the scientific journal article in scientific work? Can a discourse of scientific practices provide a superior, alternative analysis, one revealing the significant labour performed by the journal article in scientific activities, but which remains occluded by IC models? I argue that epistemological discourses of science are too weak to bear the burden of establishing the proper role and value of the journal article in science. A corollary is that they cannot demonstrate the value of collection, access, and retrieval systems, and of those who design and operate them. Discourses of scientific practice are needed for thinking about science's formal literature. When information is deflated, the way is cleared for materialist studies of documents, documentary practices, and documentary circuits in science.

4
Studies of Scientific Practices

Recent studies of scientific practices emphasize the 'patchiness' or 'motley' of the various sciences rather than the conceptual unity of a single system. The picture of 'one world, one reality, one truth' which reflects the science-as-knowledge model – a picture Hacking calls a 'metaphysical sentiment' rather than a coherent doctrine (1996, 44) – has faded after almost three decades of studies revealing wide variety among the aims, techniques, methods, instruments, concepts, theories, and many other elements involved in scientific practices. It now comes as no surprise to find *The Disunity of Science* (Galison and Stump 1996) among leading publications on contemporary philosophy of science. Nor is it unusual to find the unity of science referred to as an ideology (Hacking 1996, 37). Studies of scientific labour have emphasized diversity, difference, incommensurability, gaps, overlaps, alternatives, and disjunctions. As Pickering puts it, 'scientific culture is made up of all sorts of bits and pieces – material, social and conceptual – that stand in no necessary relation to one another' (1992, 8). The varieties of scientific practices and the complexities of scientific culture revealed by these studies bring into sharp relief the limitations of science-as-knowledge models. In contrast to the view of science as a unified conceptual field consisting in scientific statements whose propositional contents are related according to universal scientific rationality, studies of practice demonstrate that 'scientific culture is *disunified, multiple, and heterogeneous*' (1995, 70; emphasis in original).

The shift from homogeneous conceptual networks to heterogeneous material networks, from thinking to doing, and from representing the world to intervening in it (Hacking 1983), shifts the burden of science studies from abstract elements, such as information, to material elements, such as instruments, machines, actors, forms of social organization, technologies, disciplined techniques and procedures, and the like. The problem is not how the immaterial elements of conceptual networks are gathered, categorized, organized, and processed, but how the many diverse 'bits and pieces' of scientific culture are assembled and related to one another such that scientific phenomena emerge and stabilize. Different investigators emphasize different kinds of items. Hacking (1992) has a taxonomy of fifteen kinds of elements upon whose mutual accommodation the stability of scientific phenomena depends. Pickering (1995) emphasizes instruments, concepts, and disciplines. Timothy Lenoir investigates (1997) institutions.

Just as studies of scientific practices have set new problems for the analysis of the nature, growth, and stability of the sciences, so too have they set new problems for analyses of scientific documentation. Documents, especially science's formal literature, account for a significant portion of scientific labour. Not only do they figure in scientific production processes, they also become resources for further work. They belong to the heterogeneous elements comprising the motley of scientific practices. Documentary forms such as the journal article, and documentary practices such as the disciplined routines of scientific writing, together with their related technologies and institutions of organization, retrieval, and classification, are among the 'bits and pieces' of scientific work. How can the roles of documentary practices and circuits in scientific practices be understood in a way that does justice both to the labour expended upon them and to what these documents are about, in an approach to science studies that decentres the notions of information and communication? What role do documents, and the labour of producing, organizing, and deploying them, play in the assemblages or networks of elements comprising the motley of scientific practices? How do discursive 'bits and pieces' fit together with others to produce stable scientific phenomena? More generally: What is the role of documentation in scientific culture?

When analysis is guided by the diversity, motley, and patchiness of scientific practices, there is no a priori reason to assume a unified role for specific document types. In science-as-knowledge models, they do have a unified role: the communication of scientific information. When emphasis is placed on practice, there is reason to anticipate that the role of document types exhibits complexities reflecting the patchiness and variety of scientific work rather than the coherence assumed by interpreting information and communication as theoretically coherent kinds. If, as Pickering suggests, knowledge-based analytical repertoires are unable to generate an adequate understanding of scientific practices, it is unlikely that IC models can reveal the complexities of the alliances forged between specific document types and both the other discursive 'bits and pieces' of scientific work and its human, material, and social elements. This chapter reveals how the varieties of scientific practices set the stage for understanding the role of the journal article in a manner that does not hold it hostage to construals of communication and information as theoretical kinds.

The Motley of Scientific Discourses

Scientists themselves often express the metaphysical sentiment of 'one world, one reality, one truth' when speaking about science. Should their views not be privileged above those of lay persons, especially above the philosophers, sociologists, historians, and anthropologists who write and speak about science while readily admitting their ignorance of the content of the sciences they study? And, insofar as scientists' talk implies that science is a single conceptual field, does their authority as experts not lend support to the importance to analyses of scientific practices of the closely connected concepts of information gathering and processing?

Yet studies of how scientists talk about their work reveal complexities that rupture the smooth discursive regularities of the science-as-knowledge model featured so prominently in their official texts, such as research papers and Nobel lectures. Rather than offering a fast track to the most plausible view of science, the talk of experts turns out to exhibit a surprising motley, as G. Nigel Gilbert and Michael Mulkay's studies have shown: 'The degree of variability in scientists' accounts of ostensibly the same actions and beliefs is, in fact, quite remarkable. Not only do different scientists' accounts differ; not only do each scientist's accounts vary between letters, lab notes, interviews, conference proceedings, research papers, etc.; but scientists furnish quite different versions of events within a single recorded interview transcript or a single session of a taped conference discussion' (1984, 11). A closer look at these discursive labyrinths suggests that expert talk deconstructs its own image of science as a unified conceptual field.

Gilbert and Mulkay discovered two very broad kinds of discursive regularities associated with 'two contexts of linguistic production,' one formal (the experimental research paper) and the other informal (interviews with scientists). These are two distinct interpretive repertoires used by scientists to describe and make sense of their work. The first, which Gilbert and Mulkay call an 'empiricist repertoire,' expresses the metaphysical sentiment of one world, one reality, one truth. The other they call a 'contingent repertoire.' It rejects the inexorable imperatives of scientific method as the rationale for crucial experimental decisions in favour of specific and highly localized contingencies. Although incompatible as interpretations of scientific practice, both repertoires are held in reserve for deployment in appropriate contexts.[1] The scientists studied used both repertoires simultaneously in informal contexts and were not only aware of the disjunctions between them, but were happy to comment upon and explain the reasons for their differences (58–61).

The objectivity of the natural world and the dictates of scientific reasoning figure prominently in the empiricist repertoire. The literary style of the research paper represents experimental moves as forced by nature. Specific discursive techniques are mobilized so that 'the findings begin to take on an appearance of objectivity' (47): 'By adopting these kinds of linguistic features, authors construct texts in which the physical world seems regularly to speak, and sometimes to act, for itself ... [the scientist] is presented either as being forced to undertake experiments, to reach theoretical conclusions, and so on, by the unequivocal demands of the natural phenomena ... The guiding principle of this repertoire seems to be that speakers depict their actions and beliefs as a neutral medium through which empirical phenomena make themselves evident' (56). Scientific rationality is imagined as embodied in a universal and transparent method easily available to and effortlessly used in the conduct of research science: 'Methods sections ... appear to be formally constructed as if all the actions of researchers relevant to their results can be expressed as impersonal rules; as if the individual characteristics of researchers have no bearing on the production of results; as if the application of these rules to particular actions is unproblematic; and as if, therefore, the reproduction of equivalent observations can easily be obtained by any competent researcher through compliance with the rules' (54). The description of the method as universal is congruent with the idea of one world and one truth: '[the scientist] is presented ... as being rigidly constrained by invariant rules of experimental procedure which are, in turn, required by the nature of the physical world' (56).

The contingent repertoire stands in sharp contrast to the idealizations of science found in the empiricist repertoire. Appeals to 'personal antagonisms, intellectual rivalries, competition for status, clashes between strong personalities, and so on' (110) undermine justifications of action and belief couched in terms of 'experimental facts.' Scientific rationality emerges as multiple and opportunistic, shot through with social and subjective elements. The stable, objective world of the empiricist repertoire is crowded out by multiple social worlds, from whose arbitrary and unpredictable collisions emerge unexpected opportunities for experimental decisions. Gilbert and Mulkay found that the 'guiding principle' of the contingent repertoire 'is in direct opposition to that of the empiricist repertoire in that it enables speakers to depict professional actions and beliefs as outside the realm of empirical ... phenomena. When this repertoire is employed, scientists' actions are no longer depicted as generic responses to the realities of the natural world, but as the activities and judgements of specific individuals acting on the basis of their personal inclinations and particular social positions' (57).

The two discursive repertoires present a problem for IC models of science because they offer alternative accounts of the same actions. Which discursive repertoire should be privileged as the 'true' account of scientific practices? If one wants to know how specific results were achieved, to which repertoire does one turn? In the talk of scientific experts, idealizations of science as a unified conceptual domain exist side by side with depictions of an agonistic social field. The science-as-knowledge model, with its tidy notions of scientific practices as moves forced by the imperatives of universal scientific rationality, is destabilized by scientists' own talk about contingency, opportunism, unpredictability, and practical reason. The centrality of information to such models is eroded by narratives of the discordant social worlds of experimental science in which unforeseen opportunities are readily seized. Such action is not explained by orderly communication of epistemic content and its impersonal processing according to fixed rules of method. Only in the empiricist repertoire could 'information' find the stability it enjoys in IC models. The contingent repertoire provides no resources for articulating an epistemic conception of scientific information. Since both repertoires are deployed, no uncontested privilege is conferred upon 'information' by what experts have to say.

Gilbert and Mulkay make a distinction with important implications for studies of scholarly communication among scientists and for information studies generally. They explain that their method of discourse analysis 'treats participants' discourse as a topic instead of a resource.' A common assumption of studies that rely upon what participants say is that perceptions and beliefs exhibit the kind of unity characteristic of a single narrative or point of view which it is then the analyst's job to discover. Remarking upon the differences between methods used in earlier studies of science and their own, Gilbert and Mulkay note that '[p]revious approaches have been designed to use scientists' symbolic products as resources which can be assembled in various ways to tell analysts' stories about "the way that science is"' (13). But when the discourse of scientists is treated as a topic, it is found to exhibit a variability at odds with any single story about the way that science is. The idea that participants' perceptions and beliefs form a unified account of the issue under study, whether it be the nature of scientific practices or information behaviour, is revealed as an unwarranted assumption when discourse is treated as a topic. Gilbert and Mulkay's statement of their approach to science studies is a useful guide for other studies that rely upon participants' talk: 'we have approached the social world of science as a multiple reality. We have abandoned the traditional socio-logical goal of producing a single, coherent account of the patterns of action and belief in science' (188).

Treatments of discourse as a topic in many fields has revealed variability, difference, incoherencies, ruptures, and contradictions. This kind of work rejects the presuppositions of such phenomenological, mentalist, and cognitivist notions as 'mental models' or 'mental images,' 'world-pictures' or 'world-views.' Analogous to studies that assume scientists aim at producing a theoretical representation of the natural world, information studies often imagine ordinary people going about their everyday lives carrying inside them pictures of their world. And like the scientists who are imagined as struggling to enlarge and complete a representation of the natural world by seeking, processing, and adding 'information' to it, so too are ordinary people imagined as seeking 'information' to round out their internal pictures of their world, or at least their particular corners of it.[2] When studies of participants' discourse is guided by conceptions of information as a theoretical kind, they most often treat it as a resource for developing or at least contributing to a single, coherent account of behaviour participants engage in when seeking, encountering, or communicating a singular kind of thing: information.

To treat discourse about information as a topic suggests that to ask scientists about their 'information uses' invites alternative descriptions.[3] Chapter 2 showed that there are tensions in the literature of information studies between realist and nominalist tendencies. Is 'information' a theoretically coherent kind, about which knowledge may be gained by virtue of its role as raw material for the cognitive processes by means of which the natural world is represented? Or is the term merely a convenient placeholder for the various and disparate kinds of texts, documents, data, images, sounds, and many other elements that play a wide variety of roles in scientific practices? If the way that scientists talk about information reveals the same variability as their talk about science, then the theoretical coherence of 'information' and its related notions, such as 'information uses' or 'information seeking,' is eroded far beyond any capacity to provide a unified theoretical foundation for the various and disparate roles that 'information' in its nominalist sense plays in scientists' talk. The variability of scientists' talk about information cannot be shown, however, as long as research on information uses, information seeking, and the like rely upon participants' discourse as a resource for the analysts' own realist narratives.

Theory, Observation, and Experiment

Epistemological models of science feature three connected assumptions about the nature of and relationships between theory, observation, and experiment. The first is that theory construction is the primary and paradigmatic activity of science; robust theories are the best exemplars of scientific knowledge. Although there are different ways of construing the unity of science (see Hacking 1996), most prominent among them is the idea that knowledge of the natural world is unified through systematic connections between its theories. The ultimate aim of science, according to this conception of unity, is a single grand theory under which all other theories are subsumed. The second assumption is that theories are tested by observations; to test a theory is the purpose of making observations. Since a theory stands or falls by the observations it implies, making observations serves science's highest aim: theory building. The third assumption is that experiments are designed to produce observations. It is one thing to know what kind of observations support or refute a theory, but another to devise an experimental set-up to generate them. Conceiving and conducting experiments are connected to theory building through the observations they generate. Knowledge, theory, observation, and experiment line up like dominoes. Science is knowledge of the world, the world is represented by theory, observations are required to test theory, and experiments generate observations.

This simplified sketch supports the centrality of information to scientific work. Experiments are information-generating devices because observation statements convey the information required to test theories. The information produced through a particular experiment does not, however, act alone; it must be combined with the information conveyed by observation statements generated by other experiments. Information conveyed by the combined observation statements of several experiments is the raw material of scientific knowledge. In this picture, the point of studies of the communication of scientific information is to identify, enhance, and streamline the channels that convey the information required for knowledge production.

This picture of the fit between theory, observation, and experiment bears little relationship to the conduct of science. Studies of practices call into question each of its three assumptions. When the object of investigation is what scientists do rather than epistemological reconstructions of what they are thought to do, theory building turns out to be neither the primary nor the paradigmatic scientific activity. Nor is the significance of observations held hostage to theory testing. Experiments, it turns out, also exhibit a life of their own. Theory, observation, and experiment are not related in practice as imagined in science-as-knowledge models. There are many varieties of theories, observations, and experiments and different kinds of relationships between them. When overly intellectualized pictures of scientific activity yield to studies of practices, many other elements of scientific culture beyond theory, observation, and experiment come to light. They are not the only three factors, or even the most important, among the many that must be aligned with one another to produce stable scientific phenomena.

Hacking (1983) shows that the work of science does not consist in observing nature so as to represent it faithfully, but in intervening in it to produce more-or-less stable phenomena from the mutual adjustments of disparate elements. In place of the rigid, logical relations assumed to hold between theory, observation, and experiment in the intellectualized accounts of science-as-knowledge models, he offers a fluid taxonomy of ideas, things, and marks, arguing that the stability of laboratory science arises from their interplay (1992, 44). Nowhere in his taxonomy does one find observations and experiments, and insofar as theory is included, it too is not a uniform kind, but gets dispersed among background knowledge, systematic theory, topical hypotheses, and modelling of apparatus. A single theoretical claim can never be refuted by observation, as Pierre Duhem famously argued, because more than one hypothesis is required to derive an empirical deduction from theory. Hacking regards his analysis 'as an extension of Duhem's doctrine that a theory inconsistent with an observation can always be saved by modifying an auxiliary hypothesis.'[4] But Duhem remained, as Hacking points out, a 'philosopher of theory,' and his was a doctrine about thoughts – it left out 'the whole teeming world of making instruments, remaking them, making them work, and rethinking how they work.' Hacking's extension of the doctrine is materialist. It is 'a thesis about the relationships between thoughts, acts, and manufactures' (30). He emphasizes 'the matériel of an experiment' – 'the apparatus, the instruments, the substances or objects investigated' (32). His taxonomy displaces theory construction from its central position in science-as-knowledge models; theory has no special priority in the list of elements that need to be brought together. 'New instruments,' Pickering notes, 'are just as likely to issue in new stable sciences as new theory' (1992, 10). Hacking's examples show that 'much truly fundamental research precedes any relevant theory whatsoever' (1983, 158). Nancy Cartwright and Thomas Kuhn also reject such overly intellectualized views of theories; they see theory construction as the production of an assortment of models with varying degrees of precision.[5] Hacking's pluralism about theory goes much further. He notes that '[e]very single year since 1840, physics alone has used successfully more (incompatible) models of phenomena in its day-to-day business, than it used in the preceding year. The ideal end of science is not unity but absolute plethora' (1983, 218). In a thoroughly anti-representationalist vein, he argues that the relationship between theory and the world is forged by action, not by thought: our 'preserved theories and the world fit together so snugly less because we have found out how the world is than because we have tailored each to the other' (1992, 31). 'Producing theory' turns out to be shorthand for a wide range of disparate activities, not for a single activity of constructing a unified conceptual field.[6] As Rouse puts it, '[s]cientific theory provides not one kind of thing we believe, but many kinds of thing we do' (1987, 85).

Pluralism about theory is one of the important findings of the studies of scientific practices. Another is pluralism about observation. 'Theory,' Hacking remarks, 'is no more one kind of thing than observation is' (1983, 210). There are many kind of observations and they have many purposes, contrary to epistemological models of science which see in theory testing the single purpose of observation. Some observations are made for this purpose, but many are not: in 'the philosopher's conception of observation ... the life of the experimenter is spent in the making of observations which provide the data that test theory, or upon which theory is built. This kind of observation plays a relatively minor role in most experiments' (167).

A very influential conception of the relationship between observation and experiment is Popper's model of 'conjecture and refutation,' which models scientific activity as the two-step process of making theoretical conjectures followed by taking observations in attempts falsify them. This model is thoroughly epistemological, and its major failing is that few scientists can be found who perform such refutations. Most measurements are taken in the course of what Kuhn called normal science, which works within a paradigm rather than attempting to refute it. 'For all the finery of conjecture and refutation, the story imagined by Popper almost never happens. People don't make precise measurements in order to test theories' (243). Epistemological models assume that the primary purpose of all scientific activity, including making observations, is to arrive at truth, and Popper's model serves this goal by eliminating error through falsification. Knorr-Cetina's studies of scientific work lead her to the conclusion that creating stable assemblages of laboratory matériel is more important than serving truth: 'If there is a principle which seems to govern laboratory action, it is the scientist's concern with making things "work," which points to a principle of success rather than one of truth. Needless to say, to make things work – to produce results – is not identical with attempting their falsification' (1981, 4).

Pluralism about observation is closely connected with pluralism about experimentation. Once again, epistemological models simplify the variety of experiments, their purposes, and their relationships to theory. Hacking finds that a 'question posed in terms of theory and experiment is misleading because it treats theory as one rather uniform kind of thing and experiment as another' (162). Perhaps one of the most commonly accepted notions about the value of experiment is replication: because scientists seek the truth about the world, they replicate the experiments of others to detect error or to provide support. But this too rarely happens. 'Folklore says that experiments must be repeatable ... roughly speaking, no one ever repeats an experiment' (231). Like theories and observations, experiments have various purposes other than to test theories, as supposed by epistemological models. Rouse notes that 'theory and experiment are developed in response to many considerations other than just their mutual articulation. There are distinctively experimental, instrumental, or theoretical opportunities, questions, problems, and constraints to which these practices respond' (1996, 130).

When the epistemological relations thought to hold between theory, observation, and experiment are undermined by studies of practices, so too is the role of information and communication in generating new scientific results. The narrative of the scientist as primarily a thinker, hard at work processing experimental data and drawing conclusions from them in order to generate theories that faithfully represent the world, is exposed as a philosopher's fable. Knorr-Cetina writes that '"truth" and "nature" are not the only casualties of the laboratory; the observer would find it equally difficult to locate those "theories" which are so often associated with science' (1981, 4). Studies of practices imply that observers would also find it difficult to locate the information scientist's abstraction, the 'communication of information.'

How do varieties of theory, observation, and experiment help us understand the role of the journal article in scientific practices? Recent studies of experimental science show that the realist's aim of science – 'the one true theory about the universe' – badly misrepresents scientific practice. For Hacking, the situation is far worse: 'I have never believed that even makes sense' (1992, 31). If the aim of experimental laboratory science is not development of representations of nature but instead production of stability among laboratory matériel, the role of journal articles or other documentary forms consists in their contribution to such work. What science builds is not constructed from 'information' conveyed through 'channels of scientific communication.' The constructs of science are material, not ethereal. The problem is to see how the journal article fits into scientific construction zones, not how its information fits into theoretical representations of the world.

Contingency, Context, and Resource

Varieties of theory, observation, and experiment are not the only complexities to bedevil the epistemological models of science in which information is prominently featured. Studies of scientific practices show that the contingencies and local contexts of laboratory labour disperse scientific decision making over a heterogeneous field of practical reason that contrasts sharply with epistemological notions of scientific rationality. This section argues that when the reasons for the decisions scientists make in producing stable laboratory phenomena are local, contingent, and highly contextualized, scientific rationality is governed by practical reasoning of local resource selection instead of universal rules of method. When contingency and context play central roles in analyses of scientific practices, *resource* replaces information.

Knorr-Cetina's finding that laboratory action is governed by the principle of making things 'work' shifts the purpose of scientific labour from excogitating theoretical representations of the natural world to getting instruments, machines, and laboratory matériel to perform successfully, a process of assembly that Knorr-Cetina calls 'tinkering' (Knorr 1979; Knorr-Cetina 1981, 34). A resource is an element of such assemblages. In contrast to the philosopher's ideal of method as rule-governed, systematic, and universal, the selection of resources by tinkering is opportunistic, contingent, local, and idiosyncratic. A resource is therefore something quite different from epistemic content. The many and varied situations and circumstances of laboratory resource selection prevent their justifications from being reduced 'to a small number of criteria, much less to a principle of rationality.' The selections of laboratory resources, unlike the selection of information in epistemological models, are not governed by principles of scientific rationality, but are instead 'the product of the co-occurrence and interaction of factors whose impact and relevance they happen to constitute at a given time and place, i.e., of the circumstances within which the scientists operate' (Knorr-Cetina 1981, 9).

An example of the opportunism of resource selection is Knorr-Cetina's case of the choice of an item usually thought to convey 'information': a scientific paper. A laboratory researcher 'working on protein had come across a paper which mentioned the use of ferric chloride as an effective method of precipitating proteins from waste water at low temperature. In the context of an energy shortage, the use of ferric chloride struck the scientist as an excellent alternative to heat coagulation, which ... consumed a disproportionate amount of energy' (36). Protein recovery was not part of the scientist's research area. Unable to secure the protein he needed for another project, he decided to generate it himself as quickly as possible, without any intention of writing a paper on this marginal aspect of his work. His selection of a previous scientific result was an opportunistic move made in specific circumstances, one which helped simultaneously to produce the required protein, re-direct his work, and result in an unplanned paper. His selection of a specific item of 'information' – a report on the use of ferric chloride as a low-temperature protein-precipitating agent – was not governed by the decontextualized criteria of scientific method in the service of theory construction, testing, or falsification. It was chosen not because it conveyed a 'truth' put to work in drawing theoretical conclusions, but because of an energy shortage, the difficulties of obtaining protein for an unrelated project, and the availability of local laboratory maté-riel to manufacture the protein. Moreover, the intersection of local con-text and professional imperatives of academic publishing produced an unanticipated journal article.

Knorr-Cetina claims that the contingency and contextualization characteristic of this case are general features of laboratory work. Like other studies of laboratory labour, her work shows that the cognitive moves in abstract and decontextualized information space featured in science-as-knowledge accounts are missing from scientific practices. If knowledge is constructed materially, if truth and nature are absent and theories are elsewhere, then 'information' is also missing from labora-tory work. As argued in the previous chapter, to say that a scientist found information that ferric chloride is a low-temperature protein-precipitating agent is not to say anything of theoretical import. Inter-pretations of information as a theoretical kind support construals of scientific resource selections as made according to principles of scien-tific rationality. But Knorr-Cetina's study does not reveal any such 'principle of rationality' governing the selection of previously docu-mented scientific results. It implies instead that information plays no special, theoretical role in resource selection and use in scientific prac-tices. If it plays any role, it is a nominalist one: the scientist found infor-mation in the sense that he found an article reporting that ferric chloride is a low-temperature protein-precipitating agent rather than working it out for himself in the laboratory. This conclusion does not mean, however, that the role of the various and disparate things denoted by 'information' in its nominalist usage – to refer to various kinds of printed texts and illustrations, electronic documents and com-munications, bibliographical databases, conference presentations, and informal conversation with colleagues, and so on – are of no interest to analyses of scientific practices. On the contrary: the concept of infor-mation as a theoretically coherent kind acts as a barrier to investiga-tions of how a wide variety of document forms become resources for scientific labour. Attention to the practical reasoning of resource selection shifts the focus from information to document selection and use. The contingencies and locality of resources of scientific work motivate analyses of how documents fit into scientific construction zones.

The shift of analytical attention from information to resource has several implications for investigations of the role of journal articles in scientific practices. First, their selection as resources for scientific work should not be expected to conform to epistemologically idealized conceptions of scientific rationality. Instead, document selections should be expected to exhibit the general characteristics of other resource deployments: 'the selections of the research process reflect interpretations which are crystallisations of order in a local contingency space' (Knorr-Cetina 1981, 39). A corollary, as Menzel (1964) noted several decades ago, is that it is theoretically meaningless to ask for a scientist's search for 'information.' Since epistemic content is absent from scientific labour, the selection of documents from scientific literature should not be expected to be governed by the universal principles of method assumed by IC models.

Second, the idiosyncrasies of resource selection imply that resource use, including use of journal articles, is unpredictable: 'If we cannot name, once and for all, the criteria according to which scientific results are chosen or eliminated, we are unable to say which selections scientists are most likely to make. If the context of selection varies over time and place and as a function of previous selection, the rationale for scientific selection will likewise vary' (Knorr-Cetina 1981, 10). The implication for information retrieval research on the use of journal articles is that there is no predictable connection between research problems and articles retrieved. Thus there is no royal road from an 'information need' articulated in terms of research topic and the 'subject' of a document as represented in information retrieval systems. The criteria of resource selection are too idiosyncratic, local, and situated for such tidy connections.

Third, the contrast between the idealized logic of scientific rationality found in the journal article and the idiosyncrasies of resource selection presents a new problem for analysing the role of this documentary type in scientific labour. Knorr-Cetina remarks that 'natural and scientific research is in principle indetermined by the scriptures (authoritative writings) of a field, as well as by its tacit knowledge, if both are thought to represent generally available information' (1981, 40). It is indetermined by the scriptures not only because the formal literature fails to convey the 'information' pertinent to the production of knowledge at the research front, but also because such knowledge production is not governed by the methodical, rule-governed reasoning on display in the journal article. If practical decisions in a 'local contingency space' make things 'work,' what is the point and purpose of a form of discursive labour – writing journal articles – characterized by a deliberate occlusion of the contingencies and contexts of resource selections? How is this idealized discursive labour, and its primary product – the journal article – related to the 'tinkering' performed in the laboratory? When epistemological idealizations of scientific method are absent from actual laboratory practices, their reappearance in science's formal literature becomes a research problem.

Citation studies provide an example. They assume that a document's references reflect logical connections between the epistemic content of journal articles. These connections are thought to reveal the state of knowledge in specific scientific fields. Once the picture of a research front emerges from co-citation studies, it reflects an idealized world of relationships between stable units of epistemic content. The contingencies and contextualization of resource selection and use, however, imply that citation studies are primarily modes of textual analysis. They may reveal useful discursive characteristics and regularities of a specific document type, but they do not reflect the real world of research science. It is helpful to have such techniques to study the discursive products of science. But when they reflect a picture of scientific method significantly at odds with the practical reasoning of laboratory science, a question remains: How is this kind of discursive labour related to the idiosyncrasies of resource selection characteristic of scientific work?

Fourth, the value of resources for scientific labour derives from their open-ended, undefined, and unrealized capacities for further work, not from stable and fixed logical relationships. Knowledge does not circulate between centres of scientific work via logical connections between propositions. When the work of science consists in local and contextualized resource use, its extension beyond local contexts is not governed by abstract epistemological rationales. Rouse notes that the 'local laboratory site turns out to be a place where the empirical character of science is constructed through the experimenter's local, practical know-how. The resulting knowledge is extended outside the laboratory not by generalization to universal laws instantiable elsewhere, but by the adaptation of locally situated practices to new local contexts' (1987, 125). Citing an example from theoretical physics, Knorr-Cetina notes that the 'charm' model of particle physics replaced the 'colour' explanation because the former 'proved to be "of interest" to other scientists because of its perceived *unrealised capacities*, or because it inserted itself as a new *resource* in the fabric of research production in the field' (1981, 84). An implication for the journal article is that its value as a resource does not consist in the logical properties of an epistemic content assumed to be always fully present – the 'information' of IC models – but rather in a potential realized only post hoc, after its appropriation as a resource for further work. The article's utility depends not upon a presumedly inherent and always fully present epistemic content, but upon the success, interest in, and value of its use as a resource. The journal article is thus far more plastic, its 'content' far more adaptable, than is assumed by IC models.

Substitution of resource for information therefore motivates investigations of the function of the journal article as one of the many kinds of resource with unrealized capacities available to scientific practices. It is useful to think of science's formal literature more like the building materials of construction zones, selected because they can be made to fit the work at hand, than like epistemic content selected according to fixed, cognitive imperatives of theory building.[7] When information and communication are construed as theoretical kinds, the paradigmatically modernist propensities of their epistemological interpretation face powerful criticisms from studies of scientific practices. But when 'information' is deflated, analysis can shift towards investigations of the production and deployment of a wide and open-ended series of document types in many disparate scientific fields. The role of the journal article in science is best understood when epistemology gives way to studies of practices, work, and labour.

Temporality and Culture

Studies of scientific practices extend the study of contingency, context, and resource beyond laboratory resource selections to the products of the laboratory.[8] Hans-Jörg Rheinberger's investigation of experimentation in modern molecular biology shows that experimental systems 'are arrangements that allow us to create ... unprecedented events' (1997, 23). 'Local wisdom,' he concludes, 'is what characterizes the practice of the sciences' (228). Rouse captures the locality of both resources and results when he notes that 'scientific practitioners develop specific "experimental systems," using locally situated materials, protocols, techniques, and skills to produce phenomena (or their manifestations) that do not exist anywhere else' (1996, 184–5). Yet science speaks not about what is peculiar to specific times and places but about enduring phenomena of the natural world. If laboratory phenomena are peculiar to time and place, the stability of 'natural' phenomena must be constructed. This section explores the role of temporality and culture in such constructions. A temporal and cultural conception of knowledge contrasts sharply with its interpretation in epistemological models of science and suggests new directions for investigating the role of the journal article in scientific practices.

The phenomena of an 'experimental microworld' – an expression Rouse prefers to 'experiments,' to emphasize the variety of objects assembled in scientific practices – will not become 'phenomena of the natural world' if they remain artifacts of a specific laboratory.[9] They must be produced in a regular manner. This is done by gradually adapting, through standardization and regimentation, the microworld's elements and their relationships for use elsewhere. If the phenomena, the inscriptions of phenomena, the configurations of matériel, and the resources of a specific laboratory are subsequently ignored by others, or if they resist attempts to use them as resources in other contexts, they run the risk of persisting merely as curios of specific times and places. Not only are locally produced devices, instruments, reagents, catalysts, and many other kinds of objects 'black-boxed,' that is, adapted, adjusted, regulated, reproduced, and made to conform to one another for use in other contexts and in different experimental set-ups, but the researchers themselves are disciplined such that their practices at different times and places are made to conform to one another. As Rouse puts it, 'if objects and events are to be more carefully regimented, sequestered, and timed, people's actions must also be more carefully monitored and controlled' (1996, 131). The labour expended upon making microworlds more natural and 'worldly' is part of the work that makes it possible for science to speak plausibly in its official voice about phenomena of an objective, natural world.

The ways in which localized products and results can be extended to other contexts may be illustrated by contrasting the following two cases. Imagine first a tribe whose ailments are cured by the secret herbal remedies of the village shaman. Scientists visiting the village discover no mappings of herbal concoction to type of ailment. What appears to the visitors as two instances of the same ailment are treated on different occasions by different remedies. Yet the treatments work.

It also seems on other occasions that different ailments are treated by what appears to be the same concoction, which, to the scientists' puzzlement, also work. Moreover, the notions of the 'same' ailment, and the 'same' remedy, are not absent from the shaman's vocabulary. They are simply distributed in ways that the scientists do not understand.[10] The herbal remedies remain for them forever mere concoctions, because they are unable to create the conditions under which the herbal remedies effect their cures. Even worse, the visitor's culture of scientific procedure allows them no option but to conclude that the concoctions do not really 'cause' the cures. The cure and the herbs become detached because they lack the kinds of associations which, in the visitors' culture, certify attributions of a 'scientifically' validated causal connection.

The second case concerns a very familiar object. The telephone is so ubiquitous in the regions of the globe configured by modern technoscience that we experience it almost as a natural object, or quasi-natural phenomenon.[11] Perhaps paradoxically, it can be encountered in this way because it seems so clearly a product of 'scientific knowledge,' those truths about nature discovered through Bell's rigorous experimental work and congealed in this one small object. The telephone spreads far and wide, it would seem, precisely because it incorporates knowledge whose validity has not only withstood the test of time but is confirmed with each familiar ring and dial tone.[12] Yet at one time the phenomenon of carrying a human voice over a wire was an isolated and highly local laboratory phenomenon. Its success in the world beyond Bell's laboratory is due to the unpredictable and highly contingent extensions of the local configurations of Bell's laboratory matériel. Had these extensions never been made, had the phenomenon interested no one, had no labour been expended to adapt and standardize Bell's laboratory devices and to routinize his procedures in new contexts, his achievement would have remained a mere curio, useful perhaps as a toy for the family children. The work of replacing the idiosyncratic components and materials of Bell's device with widely available and standardized parts, of adapting the tentative steps involved in its laboratory creation to the imperatives of production techniques, of translating the interests of others such that their goals are achieved by its development, production, and installation from one end of the globe to the other – all this work and much more transform a local gadget into a widely dispersed, technoscientific triumph. Following Latour's 'third rule of method' (1987, 258), we can say that the telephone's success is not due to the scientific knowledge applied in its production; rather, once his device becomes a success through the labour of forging and strengthening the many associations between Bell's laboratory curio and other laboratories, factories, engineers, scientists, politicians, and cadres of workers, it becomes legitimate to *call* the knowledge leading to its production 'scientific.' Without these associations and the institutionalized practices that allow local laboratory elements to be adapted for use elsewhere, the laboratory phenomenon's scientific credentials remain weak and defenceless. Lacking such credentials, it becomes craft or folk art. Whether it is science or craft, however, does not depend upon the intrinsic characteristics of the phenomenon that emerged in Bell's laboratory. It depends instead upon its extensions to and connections with particular techniques, practices, actors, institutions, and discourses.[13]

Two important differences between these cases have implications for the nature of scientific knowledge. The first consists in the different *cultural* locations of the shaman's work-bench and Bell's laboratory. The context for the invention and development of the telephone and its transformation from laboratory curio to scientific object is a scientific culture consisting of the particular techniques, practices, actors, institutions, discourses, social relations, and the culturally extended network of associations that make it possible for the telephone to become a technoscientific achievement. The cultural context of the village shaman is very different. In the shaman's world, each concoction's uniqueness is its point and purpose. It is meant as a mystery, not as a standardized product of a production line. Its methods of creation are not embedded in disciplinary practices extended to an educated 'public.' The connections and associations between the production of an herbal remedy and the rest of tribal culture support and enhance its meaning as a unique product and the shaman's status as a master of mystery.

Scientific culture is as much a historical contingency as tribal culture. The sciences are no more a necessary feature of our world than, say, classical music. Both the natural sciences and the varieties of classical music have their historical origins, struggles of development, periods of greater and lesser creativity, heroic figures, and grand narratives. Shamanistic herbal remedies, no matter how successful in curing ailments, do not count as scientific unless the cultural contexts of their characteristic phenomena construct, maintain, and discipline their modes of adaptation, standardization, routinization, and extension in specific contexts and in fulfilment of specific goals and intentions. Given a cultural context in which herbal concoctions may be reduced, distilled, mixed, reacted, tested, monitored, analysed, represented, compared – in short, taken up in further work fully authorized as scientific by a wide variety of institutions, social relations, practices, and techniques, and situated in the relevant scientific narratives – they can become the kind of phenomena about which scientific knowledge may be gained. Otherwise, they remain a curious and even fantastic series of unique creations, at best the products of craft skill, folk art, or 'primitive' tribal practices. Even if the village remains free of the ailments afflicting its neighbours, even if the shaman cures the ailments of the long-suffering neighbours – and even if they cure ailments contracted by the visiting scientists themselves – these small successes will not be scientific triumphs as long as they are not situated within a scientific culture. Disconnected from specific institutions of adaptation, standardization, and resource production, both the shaman's concoctions and the phenomena produced in Bell's laboratory remain nonscientific curiosities. To call a phenomenon 'scientific' is to give it a *cultural* designation.

A second difference between the two cases consists in their different possibilities for extending their relevant phenomena into the future. They can therefore be said to exhibit a different *temporality*. Voice transmission over a wire, when it appears as a phenomenon of the laboratory, is one of those 'highly localised activities [that] usually make sense primarily as practices that aim to be informative about objects and events that do or might occur elsewhere' (Rouse 1996, 185). When certain elements of the laboratory mutually reinforce each other in ways that allow for adaptations to future contexts of scientific work, the practices of their creation are scientifically meaningful. They are connected to both previous and future activities by the temporal relays traced by the adaptations and extensions of mutually reinforcing local laboratory elements that produce more-or-less stable 'packages' of things, devices, discursive objects, people, social relations, routines, and techniques. Rouse notes that a research project's 'significance and standing depends upon other research continuing to develop in ways that support and build on it (or alternatively, on the course of research changing, whether anticipated or not, in ways that constructively engage it)' (187). Practices that remain disassociated from possibilities of transforming laboratory phenomena into future resources are not scientifically meaningful. Thus not only are the activities we call 'scientific' located within a scientific culture, they also exhibit a temporality traced by the opportunistic deployments of the mutually reinforcing 'packages' produced by laboratory labour. Rheinberger stresses the importance of the temporality of experimental work: 'Research systems are tinkered arrangements that are not set up for the purpose of repetitive operation but for the continuous reemergence of unexpected events. Experimentation, as a machine for making the future, has to engender unexpected events. However, it also channels them, for their significance ultimately derives from their potential to become, sooner or later, integral parts of future technical conditions. This movement implies that, in the last resort, it is the future integration into the realm of the technical that grants the scientific objects their "legitimate position" within the history of knowledge' (1997, 32–3).

By contrast, the absence of the cultural elements that would permit the isolation and definition of a specific substance, and the associations through which it can become sufficiently stabilized to be made available as a resource for further work, means that shamanistic production of herbal mixtures is not 'informative about objects and events that do or might occur elsewhere.' Much craft knowledge, like gourmet cooking or flower arranging, is not intended to be informative about such objects or events. Uniqueness is often an essential cultural characteristic of particular phenomena, intended to create distinctive objects in explicit opposition to the kind of standardization and routinization required by resource extension to new contexts. The important point is that their intrinsic characteristics may be no different from phenomena of the scientific laboratory. But unlike the latter, they are not projected into the future in the same way: they exhibit a different temporality. Once created, they stand to the future merely as singular, enduring objects, not as resources for further collective work.

The temporality and cultural character of the natural sciences have implications for the nature of scientific knowledge. Studies of scientific practices have shown that a device, inscription, or technique becomes an exemplar of scientific knowledge through procedures of standardization that permit locally adapted 'epistemic packages' to be deployed across local contexts. The multiple adaptations of such packages are what Rouse understands as 'epistemic alignments.' Commenting on the 'practices that aim to be informative about objects and events that do or might occur elsewhere,' he points out that such alignments constitute relationships from which 'knowledge' emerges as an *effect*:

The gradual adaptations, reproductions, extensions, and standardizations of these localized activities ... constitute 'epistemic alignments' ... they enable these localized activities and achievements to be informative ... Moreover, they do not merely determine whether they are informative but also what they are informative about, in what ways, and with what significance ... Some elements of ... arrays of localized projects, practices, and capabilities turn out to reinforce and strengthen one another and are taken up, extended, and reproduced in various new contexts. Others remain isolated from or in conflict with ... emergent strategies and gradually become forgotten or isolated curiosities ... I take these strategic alignments to be constitutive of what can count as knowing. (1996, 185–6)

According to this view, knowledge is an effect of synchronic and diachronic epistemic alignments that consist in the extension of experimental elements and relationships to further scientific work. The significance and scientific meaning of a research project is therefore not an intrinsic property. Unlike representational conceptions of knowledge as wholly present, the concept of epistemic alignments implies that knowledge is temporally diffused or deferred. Rouse points out that 'a statement, a skill, or a model does not acquire epistemic significance in isolation or instantaneously, depending instead on its relations to many other practices and capabilities and especially on the ways these relations are reproduced, transformed, and extended' (186). Scientific knowledge is *emergent* and incomplete, consisting in a series of temporal, open-ended, extendible situations brought about by multiple epistemic alignments: 'Knowledge (in the form of epistemic alignments) is best understood not as a system of propositions or a cognitive state but as a situation in the world. A situation is not fully determinate in the present (it is not just that we do not yet know what situation we are in but also that it is not yet determined what our situation is). In this sense, whether a body of scientific work is knowledge is metaphysically analogous to whether an event is "the decisive turning point" or whether a runner crossing the plate in the second inning is "the winning run"' (187). But unlike the baseball game whose eventual completion determines whether the second-inning run is the winner, the game of science is in principle never over: 'knowledge is understood as an aspect of ongoing engagement with the world and not as a project that could in principle be brought to completion' (194).

Such an account stands in sharp contrast to the reification of knowledge as a theoretical kind generated by epistemological commitments to the coherence of conceptual fields. Four important consequences ensue for the role of the journal article in science, each of which contest the role implied by IC models, regarding (1) the article's relationship to scientific truth; (2) the substitution of epistemic alignments for epistemic content; (3) the temporal and emergent character of information and communication; and (4) the fate of scientific rationality in epistemic alignments.

First, if the epistemic significance of a scientific product is temporally diffused and deferred, the selection and use of articles cannot depend upon their truth-values. Information and communication models assume that the value of the information conveyed through channels of scientific communication consists in the epistemic content of the transmitted messages. But if knowledge is an effect of temporally extended epistemic alignments, truth is likewise emergent, deferred, and diffused. Since the truth values of an article's statements emerge from their strategic alignments with other elements of scientific culture, they cannot account for their selection. Articles not taken up as resources move to the archive of science because of the historical contingencies of current and future research directions: 'some scientific work can become marginal to or even excluded from scientific knowledge, not because it is false, but because it does not fit in appropriately with ongoing research' (Rouse 1996, 187).[14] The picture developed by IC models of the journal article's role in knowledge production is undermined when truth is not an inherent property of epistemic content. The concept of the truth value of the propositional content of statements therefore offers no support for models of scientific work in which the communication and processing of information in the form of epistemic content play a key role.

A second consequence for the role of the journal article follows from the substitution of epistemic alignments for epistemic content. According to IC models, documents communicate information because they convey epistemic content.[15] The document form becomes insignificant, analogous to the role of the package in studies of drug trafficking, where the object of analysis is not the package but the drug. The radical point of a cultural approach to science is that *there is nothing to convey*. Rouse points out that the 'focus shifts to occasioned utterances and other performances, which are significant not as tokens of conventionally fixed meaning-types but only as mappings of various signifying acts onto others. Put another way, *there is no propositional content intervening between occasioned utterances and the myriad ways they can be taken up and used on subsequent occasions* (along with the ways they themselves appropriated previous significations)' (113; emphasis added). Paying attention to epistemic alignments 'enables us to understand the ways scientific practices sustain and reinforce one another to constitute knowledge without our having to postulate an underlying community of scientists, a scientific consensus on theoretical and methodological presuppositions, *or a multiply realizable "content" of knowledge contingently embodied in those particular practices,* (193–4; emphasis added). For information studies, the implication of the shift in focus Rouse refers to involves a change from abstract theoretical objects such as epistemic content to *documents* interpreted as different material kinds of temporally and spatially situated bundles of inscriptions embedded in specific kinds of cultural practices. Contrary to Dervin, the scientists' deployment of statements in journal articles is less like sense making than like working with *bricks* or *things* that get reshaped and refitted in building assemblages of heterogeneous elements under construction at particular times and places. The interesting questions are not about the communication of information, but about how specific statements are variously aligned to generate different 'significance-effects,' about how they are used as resources in different contexts, and about how their meaning, significance, and epistemic status emerges from their temporal extensions.

Because IC models focus on a statement's epistemic content, studies of communication among scientists and engineers have often taken the form of statistical reports of 'uses' (which, as we have seen, generally means little more than *retrievals*) of the primary discursive objects of science. But when epistemic content is replaced by epistemic alignments, what is important about a statement is what happens to it after it is constructed. Latour sees the labour of constructing a statement's significance as an agonistic process:

'A statement is ... always in jeopardy, much like the ball in a game of rugby. If no player takes it up, it simply sits on the grass. To have it move again you need an action, for someone to seize and throw it; but the throw depends in turn on the hostility, speed, deftness or tactics of the others. At any point, the ball may be interrupted, deflected or diverted by the other team – playing here the role of the dissenters –and interrupted, deflected or diverted by the players of your own team. The total movement of the ball, of a statement, of an artifact, will depend to some extent on your own action but to a much greater extent on that of a crowd over which you have little control. The construction of facts, like a game of rugby, is thus a collective process. (1987, 104)

Applying Latour's metaphor, we see the information use studies favoured by IC models as analogous to 'watching a rugby game on TV where only a phosphorescent ball was shown. All the running, the cunning, the excited players would be replaced by a meaningless zig-zagging spot' (107).

The third consequence for the journal article follows from the first two. If neither truth nor epistemic significance are inherent properties of documents, then neither is information. Whether an article is informing depends on what happens to it later on, as it becomes implicated in particular epistemic alignments. Its informing character is therefore emergent, an effect of its enrolment in further projects, rather than a consequence of the completeness and presence of epistemic content. The article communicates no information by itself, whether information be conceived as inhering in the text or in 'interpretations' resident in the minds of representing subjects. Instead, its significance depends upon the temporal, open-ended, and in principle incomplete epistemic alignments in which its inscriptions are engaged. In one of his few remarks on scientific literature, Rouse notes its fluidity: 'What is proposed as possible new knowledge, whether in informal discussion or initial publication, has an element of tentativeness about it. What is gathered together in retrospective judgment is always oriented toward a further advance and shaped by that projection. The scientific literature itself is always continually reorganizing what is known as a resource for further investigation' (192).

It follows too that communication is as much a casualty of the diffusion and deferment of scientific knowledge as truth and information. An article may be said to communicate information when its alignments with other elements of scientific culture 'make things work.' But things have not been made to work because of the communication of information. If knowledge is temporally emergent from the shared practices and labour of scientific culture, it is not 'something possessed by knowers and transmitted or exchanged through communicative interaction' (Rouse 1996, 181). But if it is not possessed, transmitted, or exchanged, knowledge production is not to be found in the communication of information.

Finally, a temporal and cultural understanding of science undermines epistemological interpretations of scientific rationality. Both synchronic and diachronic epistemic alignments offer resistance to intended extensions, reproductions, and adaptations. Stabilizing the elements of alignments is hard work. John Law points out that the alignment or 'association of unhelpful elements into self-sustaining networks that are ... able to resist dissociation' involves the labour of negotiating and resolving conflict: 'Elements in the network prove difficult to tame or difficult to hold in place. Vigilance and surveillance have to be maintained, or else the elements will fall out of line and the network will start to crumble ... there is almost always some degree of divergence between what the elements of a network would do if left to their own devices and what they are obliged, encouraged, or forced to do when they are enrolled within the network' (1990, 114). But the work of overcoming resistance is itself a source of knowledge, because it stabilizes resource extensions beyond local circumstances: 'Where there is (possible) resistance, new and more powerful techniques will be sought, more precise and careful measurement will be provided, and theoretical models will be refined to eliminate or bypass possible sources of inaccuracy or unrealistic assumption. These various refinements are themselves new knowledges and in turn often provide further new directions or problems for research. Hence, from around the specific points where knowledge is resisted emerges a whole cluster of new local capabilities and their extension into new contexts' (Rouse 1996, 192–3). Knowledge grows and develops, not according to a universal structure of scientific reason, but by the work of overcoming resistances to resource extension. In IC models of scientific knowledge production, the only form of resistance to selection and alignments of epistemic content are violations of logical and epistemological imperatives of scientific method as exercised by rational scientific subjects. But Rouse points out that the epistemic alignments actually forged in scientific practices are not governed by scientific rationality: 'The relevant forms of "resistance" to knowledge cannot easily be reduced to traditionally narrow epistemic categories. Obviously, knowledge can be resisted because there are gaps in the data, dubious assumptions in the theoretical models, countervailing evidence, or conflicts with other work. But it can also be resisted because the procedures and capabilities for its articulation and development are too expensive, environmentally unsound, cruel to animals, politically sensitive, of too little or too much interest to the military, unprofitable, and so forth' (193). In studies of practices, abstract models of rationality are replaced by investigations of alignments of material elements in historically contingent cultural and political fields. The labour involved in the production and deployment of journal articles is also located in this field.

If the significance of writing in scientific practices is best sought by tracing the role of inscriptions in temporally and culturally situated epistemic alignments, there is no a priori reason to suppose that any specific document type, such as the journal article, will emerge as a significant component of the networks of heterogeneous elements that constitute scientific culture. To information studies such as librarianship, textual editing, analytical bibliography, the history of print and publishing, document forms matter. If it should turn out that documents matter to the natural sciences, it will be because scientific labour intersects with the complex, interrelated connections between publishers, libraries, academic imperatives, and the social relations mediated by specific document forms. More specifically, it will be because there are identifiable activities among those constituting scientific labour that are configured by institutional imperatives of documentary circuits and practices. The significance of culture, contingency, and temporality suggests that, insofar as information studies have an abiding concern for documents, their intersection with science studies depends upon whether documents are significant elements of the epistemic alignments constituting scientific knowledge. Trying to satisfy its hunger for theoretically based disciplinary knowledge, information studies have moved in the direction of representation, IC models, and their associated decontextualized and dematerialized concepts of information and communication. In the process, the discipline has moved away from its erstwhile concern for document forms. But if, as I argue later, the journal article matters to scientific practices, how it matters becomes a resource for that part of information studies concerned with the documentary basis of scientific knowledge, albeit at the expense of information and communication as significant explanatory concepts.

Hacking's taxonomy of disparate elements aligned to produce the stability of the natural sciences, Rouse's epistemic alignments, and the emphasis in studies of practices on the networks of heterogeneous elements comprising scientific knowledge all suggest that the importance of the journal article is best sought by tracing the alignments of this documentary form, rather than its 'information.' A legitimate research area is therefore the identification of specific cases of significant adaptations of laboratory matériel, personnel, devices, instruments, theoretical models, techniques, social roles, and the many other heterogeneous elements of scientific culture to the imperatives of producing the formal literature of the natural sciences. What kinds of elements are aligned with specific documentary types, and which properties of these elements and types make a difference to these epistemic alignments? What kind of resistances and accommodations to and by documentary types stabilize the associations shown by contemporary science studies to be typical of the construction of scientific knowledge? In what 'epistemic packages' are specific document types to be found, and what is their role in them?

The Mangle of Practice, Disciplinary Agency, and Masses of Documents

Pickering's concept of the 'mangle of practice' elaborates in more detail the resistances and accommodations that offer a promising line of analysis for the role of the journal article in scientific practices. His view is of a 'dance of agency' in which the heterogeneous elements of epistemic alignments undergo significant alterations in their accommodation to resistance. 'Human and material agency are reciprocally and emergently intertwined' in scientific practice in such a way that each become altered in the process of becoming tuned to one another in 'a *dialectic of resistance and accommodation*' (1995, 21, 22).[16] Pickering's 'mangle' draws 'attention to the emergently intertwined delineation and reconfiguration of machinic captures and human intentions, practices, and so on' (23). Because tuning, or mangling, '*works both ways*, on human as well as nonhuman agency' (16), he calls the space of emergence in which scientific practices are located 'a *posthumanist* space, a space in which the human actors are still there but now inextricably entangled with the nonhuman, no longer at the center of the action and calling the shots. The world makes us in one and the same process as we make the world' (26; emphasis in original). If the material world of machines offered no resistance to human intentions, and if the social world of humans offered no resistance to the operations of machines, no accommodations would be required. But there is resistance on both sides. Machinic assemblages require many adjustments before they begin to work as intended (if they ever do), and often intentions, goals, and purposes have to be accommodated to what can be made to work.[17]

Like Hacking, Pickering is amenable to expanding the list of elements that figure in the mangle of practice. Although their contingency, multiplicity, and heterogeneity make it impossible to provide a final taxonomy, certain kinds of elements stand out: 'The overall view of scientific culture developed so far is one of multiplicity and hetero-geneity spanning all sorts of machines and instruments, conceptual and representational structures and systems, disciplined practices, social actors of different scales and their social relations, and so on' (111). Each of these kinds of elements may, in a specific instance, offer various sorts of resistance, and each will make various kinds of accom-modations to the resistances of others.[18] Although Pickering has little to say about scientific literature, either as a resource or as a source of resistance, his case studies raise several points that can be developed further to elucidate the role of writing in scientific practices.

The first study of interest is Giacomo Morpurgo's hunt for the free quark.[19] Although described as a mangling of conceptual practices in their accommodation to material agency, Pickering's account has impli-cations for the agency of documents in the mangle of practice. Morpurgo had to reconcile two kinds of conceptual accounts with adjustments of his experimental equipment. The first is his interpretive account of the operation of the experimental detection device (the magnetic levitation electrometer). The second is his phenomenal account, or his 'conceptu-alisations of the aspects of the material world that his experiments were designed to explore' (81). The taken-for-granted physics of his day, instantiated in its literature, offered resistances to each.

Morpurgo's interpretive account was based upon classical electro-statics. Yet at various stages of his work he had to revise or 'mangle' this account in order to accommodate the unexpected outcomes of material agency at work in his laboratory. The first of several such accommoda-tions consisted in incorporating 'an additional term in the formula for calculating charges to allow for electric field gradients acting on [his target material]' (80). The need to tinker with a law of classical electro-statics in this way arose from the disjunction between it and the workings of his apparatus. His conceptual tinkering, however, faced resistance: 'Clearly he did not explore [the indefinite space of possible electrostatic interpretive accounts] at random. Not just any mathemati-cal formula could count – for him or his colleagues – as a plausible interpretive account ... Morpurgo sought to tie the particular interpre-tive accounts that he fixed upon *into the general frame of established knowl-edge in electrostatics and prior applications thereof* (80; emphasis added).

A similar form of resistance greeted Morpurgo's accommodations of his phenomenal accounts to another set of unexpected outcomes of material agency. His aim, Pickering explains, was to forge 'an *associa-*

tion between the material and conceptual elements of his practice that would amount to one of *translation*, in which his interpretive account would translate the performance of his material instrument into one of his pair of phenomenal accounts' (81; emphasis in original). The problem Morpurgo encountered was yet another set of 'puzzling material performances' (82). His 'observations ... seemed ... to point to the conclusion that electric charge was continuously divisible' (83). But classical electrostatic theory maintains that all charges are integral, quantized in units of e, the charge on the electron. Following Murray Gell-Mann's particle physics in his hunt for the free quark, Morpurgo was searching for a particle with third-integral charges, but which had never been found in isolation. His observations could be reconciled neither with the discovery of free quarks with third-integral charges nor with classical electrostatic theory, which rejected continuously divisible electric charge. Thus for Morpurgo, integral charges or third-integral charges *were the only possible phenomenal accounts*. The reason, as Pickering explains, is that although Morpurgo 'found it conceivable that the odd, rare quark might exist ... *he had not embarked on his program of experiments in the expectation of finding that standard lore on the quantization of charges was entirely misconceived*' (83; emphasis added). In this case, as in the case of tinkering with his interpretive account, classical electrostatics offers resistance to particular manglings. Morpurgo could – and finally did – accept an accommodation that tied the workings of his machine and his adjusted interpretive account with a phenomenal account that charges are integral, but at a price: he gave up his hunt for free quarks rather than accept a phenomenal account based upon continuously divisible electric charge. To accept the latter would require a massive mangling of classical electrostatics, a labour that Morpurgo was not prepared to undertake.

Pickering's account of the hunt for the free quark attributes agency to 'conceptual practice' in the form of 'established knowledge' and 'standard lore.' But what is a 'conceptual practice'? According to epistemological models of scientific practices, established electrostatic theory is a way of representing part of the world. The resistance of established physics, according to such a view, is grounded in scientific truth expressed by established theory: Morpurgo did not accommodate his conceptual practice to his observations of continuously divisible charge because to do so would be to claim as false what was known to be true. Epistemological models interpret conceptual practice as thought governed by principles of scientific reason.

Epistemological models reflect IC models of formal scientific literature. The journal article supports the resistance offered by established conceptual practice to radical tinkering because it conveys or communicates 'information' in the form of the epistemic content of its propositions. Morpurgo could not accept that electric charge is continuously divisible because the proposition that the charge carried by 'free' fundamental particles is integral is central to classical electrostatics. Gell-Mann's physics includes the proposition that the charge carried by quarks is third-integral. The proposition that electric charge could be continuously divisible is incompatible with both established electrostatics and Gell-Man's physics. The formal literature supports the resistance offered by established physics to Morpurgo's observations of continuously divisible charge because it conveys the propositions of these two branches of established physics.

Studies of scientific practices suggest another role for the literature. When occasioned utterances or inscriptions replace epistemic content, Morpurgo faces neither truth about the world nor its representation by the theories of physics. Instead, he faces a mass of documents. The occasioned inscriptions of third-integral quark charges and integral free-particle charges in the 'standard lore' or 'scriptures' of science have such strong associations with other occasioned inscriptions, and are deployed in such deeply routinized and standardized suites of disciplined and disciplinary activities, that they resist attempts to undo them by local laboratory mangles that tie together tinkerings of equipment with a phenomenal account of continuously divisible electric charge. It was easier for Morpurgo to tinker once more with his equipment, thereby eliminating fractional charges altogether and thus bandoning his hunt for the free quark, than to undergo a thorough mangling, by science's currently authorized inscriptions, of the statements he would be required to submit to the institutionally authorized platforms of enunciation of scientific truth. Morpurgo's decision to abide by classical electrostatics is not a case of following the dictates of scientific rationality, nor of knowing the limits of one's abilities, but of an acute awareness of a statement's probability of survival when at odds with masses of authorized inscriptions. When he follows others in declining to challenge classical electrostatics he strengthens the network of associations holding it in place. There is nothing about the world that would prevent him from producing an accommodation that favoured his observation of continuously divisible charge. The truth of the world, as Hacking insists, explains nothing.

When scientific knowledge is seen as emergent, deferred, and diffused, conceptual and representational structures may be materialized as networks of occasioned inscriptions of masses of documents. In the remainder of this section, I use some ideas from Wittgenstein and Foucault to elaborate this suggestion and argue for it. My motivation arises in part from my sense that Pickering's failure to fully materialize his 'conceptual structures' comes close in more than one place to falling afoul of the metaphysical traps of hidden forces and the agency of abstract entities that he otherwise seeks to avoid. In seeking to include conceptual practices in the work of science, he on occasion seems to reify concepts and conceptual structures. Yet his notion of the mangle of practice has great value for understanding the role of writing in scientific practices.

Pickering uses a machine metaphor to explain how conceptual practices exhibit disciplinary agency: 'Conceptual structures, one can say, relate to disciplinary agency much as do machines to material agency. Once one begins to tinker with the former, just as with the latter, one has to find out in practice how the resulting conceptual machinery will perform. It is precisely in this respect that dialectics of resistance and accommodation can arise in conceptual practice' (117). Talk about conceptual machinery can easily lead to the idea that correct and incorrect conceptual moves are determined by the stability of the concepts themselves, as Wittgenstein has shown. For example, conceptual moves in arithmetic may be seen as determined by the concept of addition, or the meaning of the symbol '+.' (It makes no difference whether the meaning of '+' is in the mind, a Platonic form, or in the text.) Talk about machine-like conceptual moves can suggest that the reason three plus four equals seven is that given the meanings of '3,' '4,' '+,' and '=,' the only possible completion of '3 + 4 =' is '7.' As the next move after '3 + 4 =', '7' is determined by the concepts these symbols represent. To borrow Wittgenstein's metaphor, it is as if each symbol were but one face of a hidden block.[20] When the symbols are arranged in the manner represented by '3 + 4 =,' one and only one block can be made to fit the space created by the combination of blocks. Conceptual machinery of this sort operates in a timeless, eternal, and non-human space.

The connection Pickering makes between conceptual structures and disciplinary agency is meant to oppose this view. He recognizes that every extension of conceptual practices to construct new scientific results involves mechanical, machine-like *transcriptions* of previous conceptual practices. Such transcriptions are analogous to writing down '7' after '3 + 4 =.' But we do not write '7' because we grasp the meaning of the symbols. Instead, writing down '7' after '3 + 4 =' in the appropriate circumstances (not in fits of madness, nor upon communicating with the dead in a séance, etc.) is the criterion of grasping the meaning of the symbols, not the result of grasping their meaning. Pickering accepts the Wittgensteinian view that practice, not mental representation, is the bedrock upon which we build our familiar symbolic edifices such as meanings, concepts, epistemic content, and propositions. Our immediate and unhesitating action (given the appropriate training) may seem to follow from the operations of conceptual machinery only because the uses of arithmetical symbols have been so thoroughly disciplined as to seem natural. Once *we* are disciplined, the 'discipline' of arithmetic appears to exercise an agency of its own. But its agency rests upon culturally and historically contingent practices that are so routinely followed that logical necessity is experienced as an inexorable force. Following Lynch's appropriation of Wittgenstein's remarks on 'following a rule,' Pickering explains his concept of disciplinary agency in terms of such routinized practices:

> Just as in arithmetic one completes '3 + 4 =' by writing '7' without hesitation, so in algebra one automatically multiplies out '$a(b + c)$' as '$ab + ac$.' Conceptual systems, then, hang together with specific disciplined patterns of human agency, particular, routinized ways of connecting marks and symbols with one another. Such disciplines – acquired in training and refined in use – carry human conceptual practices along, as it were, independent of individual wishes and intents. The scientist is, in this sense, passive in disciplined conceptual practice ... I want to redescribe this human passivity in terms of a notion of *disciplinary agency*. It is, I shall say, the agency of a discipline – elementary algebra, for example – that leads us through a series of manipulations within an established conceptual system. (115)

The mangle of practice traces a route from 'conceptual and representational structures and systems' through the machine-like *forced moves* of conceptual practices to *'detached disciplines'* (101–12) and to a disciplinary agency experienced as the independent and timeless operation of conceptual machinery. But if it is not the meaning of a symbol that determines its correct and incorrect use, what, then, does? Pickering appropriates the Wittgensteinian answer: *nothing* determines it. *Every* extension, even writing '7' after '3 + 4 =,' is in principle a *free move*. Similarly, every extension of scientific culture to create new knowledge is also at bottom a free move.[21] The possibility of a *correct* move in arithmetic, or, in the case of the conceptual practices of science, a scientifically valid or rational move, depends not upon a necessity but a contingency: that humans do in fact, often with explicit training, deploy symbols in routinized, public practices. It is because they can do so that some of these practices may be disciplined to the point that only particular moves count as correct. It is a brute fact, a contingent feature of the human world, that there are such patterns of behaviour we call 'going on in the same way,' and that so much importance is attached to them. Some of these practices, or patterns of behaviour, are more or less plastic, flexible, and alterable without repercussion (Wittgenstein notes, for example, that sometimes we alter the rules of a game as we go along).[22] But many are not, including those central to various aspects of our culture and, if Wittgenstein is right, to our logic. To ensure the survival and integrity of important practices, or goings-on-in-the-same-way, disciplinary structures are institutionalized and enforced in diverse ways, as Foucault has shown: classification systems, specific representations such as charts, graphs, tables, and so forth, regimes of education and knowledge (i.e., pedagogical and scholarly 'disciplines'), and even architecture. Such institutions ensure that patterned behaviours are policed, regulated, standardized and routinized, made the object of training, rehearsal, education, drill, and exercise. Thus disciplinary agency not only offers resistance to cultural extensions by defining what is incorrect, wrong, and deviant, but is also productive, enabling cultural extensions by defining what is appropriate, proper, warranted, correct, and rational.

Pickering emphasizes that every extension of conceptual practices to construct new scientific results involves mechanical, machine-like *transcriptions* of previously established conceptual practices, analogous to routinely following '3 + 4 =' by '7.' But in the cultural extensions peculiar to the conceptual practices of science, the point of such transcriptions is to provide a point of departure for free moves that lead research in new directions: they are 'goings-on' in ways *different* from familiar routines. In Pickering's terms, they are moves attributable to human rather than disciplinary agency. In conceptual as in material practices, according to Pickering, scientific labour is a dance of agency. The new results generated from the dialectic of resistance and accommodation in conceptual practices arise from both the resistances and the accommodations of disciplinary agency.

Sir William Rowan Hamilton's construction of quaternions provides an example (see Pickering 1995, 126–38). At one point in his work with two geometrical interpretations of the imaginary numbers i and j, both equal to $\sqrt{-1}$, a 'dramatic and eventually ... far-reaching move ... struck Hamilton as possible. It was to abandon the assumption of commutation between i and the new square root of -1, j. In ordinary algebra, this assumption – which is to say that $ab = ba$ – was routine. Hamilton entertained the possibility, instead, that $ij = -ji$' (132). For Hamilton's innovation to become part of science, it must be translated into disciplinary agency. It must become embedded in new practices that flow as smoothly and as routinized as the previously established practices from which it emerged. Hamilton's success depends upon others taking up his innovation in their own scientific work, in 'going on' from ij to $-ji$ in new practices as naturally and unhesitatingly as 'going on' from '3 + 4 =' to '7.' His challenge is something like getting others to play a new game. If the game is uninteresting, useless, too complicated, or simply so bizarre that no one but its inventor can claim to know what it is to 'go on in the same way,' it will be ignored and rejected. But if Hamilton can connect his innovation to previous work, if he can insert it into scientific culture as a productive resource for others, then it stands a good chance of becoming a routinized practice and certified as a new scientific result.

Various factors play a role in the success or failure of Hamilton's innovation and others like it. What is the role of formal writing in routinizing such moves? Pickering offers few clues to situate documentary circuits among the 'regularized, routinized, standardized, [and] disciplined human practices' (102) of scientific work. Although he interprets disciplinary agency not as the product of private, mental concept manipulation, but constituted instead by a set of overt practices involving public uses of utterances and inscriptions, his presentation of Hamilton's discovery of quaternions is a narrative of a solo performance. To arrive at the role of documentary practices in the mangle of Hamilton's conceptual practice, Pickering's notion of 'disciplinary agency' needs to be more thoroughly materialized.

The routinization of new conceptual moves exemplified by Hamilton's construction of quaterions can be materialized by Foucault's notion of the micropractices of disciplinary power. Foucault argues that power is sustained by and relayed through many interconnected, minor routines. In many cases these routines are maintained by a wide variety of interlocking, disciplined deployments of many different kinds of documents. His remark in *Discipline and Punish* (1979) on the role of documents in the various 'examinations' of individuals that underlie surveillance procedures (e.g., medical examinations, and examinations given to students in schools) is highly suggestive for understanding the role of scientific literature: 'The examination that places individuals in a field of surveillance also situates them in a network of writing; it engages them in a whole mass of documents that capture and fix them. The procedures of examination were accompanied at the same time by a system of intense registration and documentary accumulation. A "power of writing" was constituted as an essential part in the mechanisms of discipline' (189).

Foucault's point is neither that the written record of the many systems of surveillance peculiar to the redistribution of 'the economy of punishment' (7) included many new kinds of documents, nor that the late eighteenth-century revolution in penal justice, in its reliance on examinations, saw a steep rise in the number of documents and a data collection apparatus that together produced an 'information explosion' of the age. He does not argue that changes in the kinds and quantity of documents were caused by a steep rise in the amount of 'information' about individuals generated by 'the examination that places them in a field of surveillance.' His point grants a far more significant role to documentation than mere communication of information. Individuals become placed in a field of surveillance by means of examinations that belong to an apparatus of 'disciplinary writing' that accumulated and organized specific characteristics and data points in various document forms which circulated in specific, highly regimented institutional circuits. The historically significant effects of such documentary practices extend far beyond conveying information: they are not communicative, but constitutive. Beyond inaugurating 'a whole series of codes of disciplinary individuality that made it possible to transcribe, by means of homogenization the individual features established by the examination' (189), the 'other innovations of disciplinary writing concerned the correlation of these elements, the accumulation of documents, their seriation, the organization of comparative fields making it possible to classify, to form categories, to determine averages, to fix norms' (190). The 'power of writing' that 'captures and fixes' individuals is not the power of simply transcribing pre-existing, objective individual characteristics into written form, thus permitting 'information' about individuals to be communicated; it is rather 'an essential part of the mechanisms of discipline,' or the disciplinary apparatuses through which individuals are constructed as objects of knowledge. Foucault's

interest is not in documentation as a means of communication of information, but as a relay of generative and formative power, through
which knowable individuals are constituted: 'Thanks to the whole
apparatus of writing that accompanied it, the examination opened up
two correlative possibilities: firstly, *the constitution of the individual as a
describable, analysable object* ... in order to maintain him in his individual
features, in his particular evolution, in his own aptitudes or abilities,
under the gaze of a permanent corpus of knowledge; and, secondly,
the constitution of a comparative system that made possible the measurement of overall phenomena, the description of groups, the charac-
terization of collective facts, the calculation of the gaps between
individuals, their distribution in a given "population"' (190; emphasis
added).

Foucault's triplet of examined individuals, the field of surveillance,
and the 'mass of documents' can be mapped onto the triplet of epistemic alignments, science, and scientific literature. The members of the
first set are related as follows: individuals are placed into a field of surveillance, and thus in a field of knowledge, by *documented examinations*.
The 'mass of documents' is an essential part of the disciplinary apparatus that constitutes individuals as knowable, because it is not the
examination alone, but the 'examination, surrounded by all its documentary techniques, [which] makes each individual a "case"' (191).
The members of the second set may be arranged analogously: epistemic alignments of heterogeneous elements are placed into a field of
science – as opposed to fields of art, craft, or folk knowledge – by a specific form of scrutiny that likewise depends upon particular 'documentary techniques.' Scientific culture is characterized by processes of
scrutiny of local laboratory arrangements analogous to the 'examination of individuals.' These processes engage the phenomena of local
laboratory arrangements in 'a whole mass of documents that capture
and fix them.'[23]

Specific kinds of documents, documentary practices, and documentary circuits are therefore central to Foucault's studies of the emergence
of 'the sciences of man' which constitute the modern 'individual'
as a theoretical object about which knowledge may be gained. The
construction of natural-scientific phenomena also depends on documents, documentary practices, and documentary circuits. The project
of constituting local laboratory phenomena as knowable things or
events of an objective, natural world depends upon documentary practices analogous to Foucault's 'examination.' Historically and culturally

contingent and institutionally authorized documentary types inscribe *forms of description* by means of which the local can be narratized as universal.

In both cases – the 'sciences of man' and the physical sciences – knowledge depends upon sites of emergence and deployments of specific kinds of discursive resources inscribed in documents. But these resources do not belong to the same ontological category as immaterial and abstract representational entities, such as propositions, concepts, mental images, and epistemic content. The discursive resources deployed to narratize local laboratory arrangements as universal belong to the category of *statements*. But statements, Foucault argues, are material: 'for a sequence of linguistic elements to be regarded and analysed as a statement ... it must have a material existence.' This materiality takes a particular form. It does not consist simply in the statement's existence at a particular time and place but in the degree of its institutional embeddedness: 'The rule of materiality that statements necessarily obey is ... *of the order of the institution* [emphasis added] rather than spatio-temporal localization; it defines *possibilities of reinscription and transcription* [emphasis in original] (but also thresholds and limits), rather than limited and perishable individualities' (1972, 100, 103). Thus its materiality is a function of institutionalized routines of significant repetition that establish relations between statements. The materiality of a statement consists in institutionalized patterns of occasioned inscription:

> This repeatable materiality ... reveals the statement as a specific and paradoxical object, but also as one of those objects that men produce, manipulate, use, transform, exchange, combine, decompose, and re-compose, and possibly destroy. Instead of being something said once and for all ... the statement, as it emerges in its materiality, appears with a status, enters various networks and various fields of use, is subjected to transferences or modifications, is integrated into operations and strategies in which its identity is maintained or effaced. Thus the statement circulates, is used, disappears, allows or prevents the realization of a desire, serves or resists various interests, participates in challenge and struggle, and becomes a theme of appropriation or rivalry. (105)

The documents that inscribe local laboratory phenomena in statements that conform to science's cultural imperatives of objectivity and universality are regulated according to authorized procedures and rules of writing, and circulated in highly routinized, standardized, and authorized circuits. The availability of authorised platforms of enunciation of scientific truth depend upon specific documentary practices. Absent the labour of defining, constructing, maintaining, and defending the authority of a specific form of writing as the primary discursive site at which local laboratory phenomena are narratized as universal, eternal, and timeless, discourses of objectivity and universality hold no sway, command no one's attention, guide no one's practices, and have no jurisdiction.

The authorized literature of science consists in sets of documentary forms that together with their authorized deployments situate local laboratory arrangements in cultural fields already understood as scientific. The scrutiny or 'examination' of local laboratory arrangements that constitute them as phenomena that belong to an already documented history of objective, universal, scientific truth takes the form of a process of writing. In Foucault's terms, the discursive site, or the authorized platform of enunciation of objective, scientific truth, is constituted by a 'power of writing.' Since the formal literature of science is an essential part of the process through which local laboratory phenomena are constructed as objects of scientific knowledge, it is therefore constitutive, not a vehicle for the 'communication of information.' The primary literature of science is its journal literature. The journal article is therefore the primary site of the 'power of writing' that 'constitutes an essential part in the mechanisms of discipline' peculiar to and definitive of scientific conceptual practices. The possibility of the enunciation of truth through scientific statements depends upon institutionalized, disciplined routines governing the production and circulation of journal articles. In Latour's terms, the journal article is an obligatory passage point for the translation of local, contingent, and temporal arrangements of heterogeneous elements of the laboratory into universal, scientific truth.

Foucault's notions of a statement's materiality in terms of its institutionalized possibilities of reinscription and transcription, and the documentary regimes that contribute to establishing and maintaining authorized platforms of enunciation for truths and knowledge claims, can be used to materialize Pickering's concept of 'disciplinary agency' and his account of the resistances of conceptual practices. If Hamilton's new, innovative, free move, which violates the law of commutation – an elementary law of algebra – is to be embraced by scientific culture, it must figure in new practices narratized as continuous with a 'surface of emergence' already culturally constituted as scientific. But – and this is Foucault's point – the required narratization is a process of writing. It is the labour of establishing connections between Hamilton's new statement, that is, that $ij = -ji$, and other statements that already occupy the appropriately authorized platform of enunciation of scientific truth. This platform is in large part constituted by a particular documentary regime, or 'mass of documents.' It is a material platform in the Foucauldian sense, one whose materiality consists in institutionalized possibilities of reinscription and transcription of Hamilton's new statement. If Hamilton's statement submitted to no imperative to forge relationships with other authorized statements of mathematical practices but was confined to his notebooks, he could 'go on' in any way he pleased. In the absence of *practices*, any extension of statements taken from previous work could, as Wittgenstein has shown in his remarks on following a rule (1958, §198 & ff.), be seen as a legitimate extension of mathematical practice. Since the materiality of Hamilton's statement is institutional, the agency it can exercise to resist deformation or thorough extinction depends upon its institutional location and associated documentary circuits. Notebook scribbles taken up and deployed in documentary regimes marginalized by the authoritative scientific culture are not certified as scientific, even if, considered from an abstract, conceptual point of view divorced from all material, historical, and cultural practices, the extensions to conceptual practice they propose are as legitimate as any other. The fate of Hamilton's innovation is thus held hostage to institutionally authorized possibilities of reinscription and transcription peculiar to the appropriate platform of enunciation and its associated documentary circuits. If Hamilton's new statement is to have a chance of being taken up in such a way as to become a new scientific result, it must be inscribed in the science's primary literature and taken up as a resource for others – it must have an institutional documentary life.

Foucault's work therefore suggests that to locate new statements in the official, canonical documents – the 'scriptures' – upon which the enunciative function of scientific truth depends, is part of the routinization of extensions such as those proposed by Hamilton. According to Pickering, such routinizations are typical of scientific work. A text does not belong to the scriptures because its content is holy; rather, its content is holy because it belongs to the scriptures. Foucault's work also suggests that it is not a matter of indifference – as supposed by IC models favouring decontextualized, abstract, and representational concepts of information and epistemic content – just how, where, when, in what form, and in what medium the statements of cultural extension are made. Finally, Foucault's work suggests that research on the roles of documentary practices among the many institutionally disciplined micopractices upon which the routinization of new, scientific conceptual practices depends can reveal the significance of writing and document forms to knowledge production.

Beyond the Laboratory

The 'patchiness' of the natural sciences, reflected in the motley of scientists' discourse, theory, observation, and experiment, and in the contingencies and contexts of laboratory work, extends well beyond the walls of the laboratory. The actant network theory of Latour and his collaborators (Latour 1983; 1987; 1988; 1992; 1993; 1996; Callon, Law, and Rip 1986; Callon 1986; 1990; Law 1990) is perhaps the most emphatic, among the many other sources of 'patchiness' discussed so far, about the thorough and complex intertwinings of elements usually considered as 'interior' and 'exterior' to the sciences. Like other writers on scientific practices, Latour et al. accept that the stability of a scientific phenomenon consists in the strength of the contingent, unpredictable, and temporary associations forged between a large number of disparate elements brought together to form a network that can withstand efforts to dismantle it, and that is sufficiently rigid to be used in projects beyond those of its origin. The importance of their approach to the issues discussed in this chapter consists in three points: (1) the dissolution, through the notion of *interessement*, of the commonplace distinction between internal and external factors in science; (2) the claim that all such distinctions are the effects of network stabilizations; and (3) the importance of connecting laboratory work to large-scale social macrostructures. All three of these attacks on 'internalism,' or the view that there is a cognitively pure *interior* to scientific work, help displace information and communication from the central role they enjoy in IC models, by greatly expanding the motley of actors (or, as Latour and his collaborators would put it, *actants*) included in scientific networks.

Interessement is the strategy of building and strengthening the associations constituting scientific networks. Law defines it as the 'action of interesting, enrolling or translating which involves one entity attracting a second by coming between that entity and a third' (Callon et al.

1986, xvii). An example is enrolling others by translating their interests into those of one's own laboratory. The stability of laboratory results is thereby enhanced, through an increase in the strength of their associations. Latour et al. show that the elements of the networks on whose strength of association the stability of a scientific fact depends are not specifiable in advance. It is never possible to predict whose interests and which strategies of *interessement* will prevail. The alignment of interests is decisive, not an imagined purely scientific interior free from external contamination. Elements of networks include groups of all kinds (academic, labour, professional, political), individuals, communities, and other social actors (e.g., corporations, governments, unions, activists, police, managers), natural and scientific artifacts including machines and instruments, ideologies, propaganda, belief systems, and discourses of various kinds including those of science, culture, technology, and politics. These are all called 'actants' since, as a point of methodological rigour, Latour insists upon treating them all as agents.[24] There are similarities to Hacking's ideas here, but Latour greatly expands the number of elements that must be coordinated, a point Hacking acknowledges: 'I have said nothing about the most important ingredient of an experiment, namely, the experimenters, their negotiations, their communications, their milieu, the very building in which they work or the institution that foots the bills. I have said nothing of authors, authority, and audience. In short, nothing of what Latour indicates by his titles *Science in Action* and *Laboratory Life'* (1992, 51).

A scientist's education, training, and socialization into a professional life that increasingly demands skill in negotiating disparate institutional imperatives also widens the context of the non-cognitive contingencies of scientific resource selections and the associations they create to include engagements with professional societies and publishing houses, the exigencies of grant management and laboratory supervision, the need to recruit and develop cadres of future scientific workers, the burdens of university administration, and the politics of government and corporate support. Far from presenting merely 'external' factors whose significance is traced via notions of 'impact' and 'influence' on the true, cognitive interior of science, these wider contexts are seamlessly connected to local laboratory contexts through the interests of disparate sets of agents. They increase the scale of the situations in which resource selections and associations are made, but as elements of networks upon which the stability of scientific facts depend, they do not represent a difference of kind from the resources found inside laboratory walls.

As a further point of method, Latour insists that the distinctions between the 'social,' the 'scientific,' the 'technological,' or the 'discursive' are the *products* or *effects* of network stabilization, and not their cause or antecedents: 'The question for us who shadow scientists is not to *decide* which one of these links is "social" and which one is "scientific," the question for us, as well as for those we follow, is only this: "which one of these links will hold and which will break apart?"' (1987, 176). The unpredictability of the kinds of actants enrolled in networks is therefore due not simply to there being so many candidates with potential interests in laboratory work. It is also due to there being no way to specify in advance the kinds of associations required to stabilize scientific facts. It is only after the agonistic process of enrolling others is successful, according to Latour, that we can say whether the link is of one kind or the other. Thus in Latour's hands, the unity of the natural sciences suffers a fate worse than their construction from disparate elements. They are disunified because the elements upon whose strength of association the stability of scientific facts depend lack determination before the association is achieved.

For Latour, the stability of facts also depends upon effecting large-scale social transformations, a point emphasized in his studies of Louis Pasteur (1983; 1988). Explaining how laboratory results can be extended outside the laboratory, Latour writes: 'In spite of all the niceties written by epistemologists on that point, the answer is simple: only by extending the laboratory itself ... [Pasteur's] vaccination can work only on the condition that the farm chosen in the village of Pouilly le Fort for the field trial be in some crucial respects transformed according to the prescriptions of Pasteur's laboratory' (1983, 115). The success of Pasteur's cure for anthrax depended upon a general condition for the success of all laboratory results: transformations of society to meet conditions of the laboratory. Thus the orchestration of the heterogeneous elements crucial to the stability of facts involves not only the laboratory items of Hacking's taxonomy, but also features of the extended environment. The studies conducted by Latour and his collaborators draw attention to the range and scale of the associations that constitute contemporary technoscience.

Although the resources required to 'do science' are unpredictable, contingent, and located in disparate contexts, they are inseparable from scientists' concepts and ideas. In her study of the world of particle physicists, Traweek uncovers a complex motley of social roles, presenting 'in detail the spaces, artifacts, and roles of the actors in the [high energy physics] community ... [and] the activities that are important in the lives of physicists and of national physics communities' (1988, x). She describes 'their social organization, developmental cycle, cosmology, and material culture' (157). Chief among the structures that elicit specific forms of action and discourage others are the physical environments of the high energy physics community, including their laboratory spaces; the experimental equipment, or devices (detectors) used to conduct sophisticated experiments of events at the subatomic level; the pattern of career development, or the rites of passage from 'undergraduate, graduate student, postdoctoral research associate, group member, group leader, laboratory director, and science statesman' (xi); the social structure of the international physics community; and the strategies that must be deployed to ensure the highest quality equipment for research laboratories. Each of these structures impose imperatives that together constitute the shared scientific world of high energy physicists, a world in which resources are selected, rejected, and deployed. Yet the factors Traweek studies are not 'external' to scientific work, since, as her study shows, the social reality of the high energy physics community cannot be separated from their theories of time and space.[25]

Scientific practices are therefore multiply complex: the motley of elements required to produce stable experimental laboratory results are embedded in and intertwined with, among other things, disparate roles, interests, actors, institutions. Scientific culture, as Pickering said, is made up of all sorts of bits and pieces. Hacking points to a motley of laboratory elements; Knorr-Cetina points to a motley of circumstances, contexts, and resources of scientific micropractices. Rouse points to epistemic alignments and Pickering to the mangle of practice. Latour emphasizes the strength of associations that include social macrostructures. Traweek points to a motley of social, material, and conceptual elements that together constitute a scientific culture. Their studies indicate that scientific labour fails to run along the smooth rails constructed by epistemological models. They demonstrate the simplicity of views of science in which information seeking and the communication of information are privileged. If, paraphrasing Hacking, doing science is better understood as intervening in the world rather than representing it, the gathering, processing, and communication of information so central to IC models gives way to descriptions of the labour involved in forging the kinds of associations of disparate elements featured in studies of scientific practices. Grounds for rejecting IC models are also grounds for rejecting the significance of the communication of information in scientific work under any meaningful interpretation of these two concepts as theoretically coherent kinds.

Writing is important to the work of science. If that work is far more complex than seeking, finding, communicating, and processing scientific information, then the concepts of information and communication are too weak to explain the role of the science's formal literature. As one of the motley of elements comprising scientific practices, the article is as much a resource as the many others featured in studies of scientific practices. The selection of journal articles and their deployments, even the very existence of institutional imperatives governing scientific work deriving from a system of scientific journals and other documents, can be fully understood only against the background of the contexts and contingencies of scientific resources generally. Epistemological models are of little use here. In 1987, Latour wrote: 'What I propose, here, as a ... rule of method, is in effect a *moratorium* on cognitive explanations of science and technology! I'd be tempted to propose a ten-year moratorium' (247). Perhaps it is time to plead for an extension.

5

Literary Technologies of Science

In the previous chapter I argued that the shift in science studies from representing to intervening and from conceptual to material networks suggests materialist and non-cognitivist approaches to the study of scientific documentation. The question, 'What is the role of documents in the material practices of science?' now directs attention to the role of a wide variety of document forms in the 'motley' of scientific practices. When the 'information' in science information systems is understood materially, as a documented statement in the Foucauldian sense and not as epistemic content of propositions, it becomes one of the 'bits and pieces' scientists use to construct stable phenomena. Studies of 'scientific information uses' or 'information seeking among scientists' are unhelpful here because they are guided by assumptions about 'information uses' as instances of understanding. A cognitivist criterion of identity for what counts as 'uses' supports their quantification, thereby providing theoretical justification for statistical investigations of retrieval satisfactions. Such studies, however, idealize the role of documents in scientific practices.

Scientific writing is an important, time-consuming, and highly institutionalized labour process in the culture of modern science.[1] Journal use studies show that scientists not only read articles, but that they spend a good deal of their time doing so. Latour and Woolgar's picture of laboratory life reveals that journal articles not only litter the workspace, but also are the topic of most laboratory discussions (1986, 52–3). Why? What role do they play? If they do not convey the information needed at the research front, and if they are not used exclusively in the production of other journal articles, what resource value do they have? This chapter addresses the question of how writing has been interpreted in recent science studies and extracts from those studies some resources for a materialist analysis of science's formal literature.

A common theme of many studies of scientific practices is that scientific writing, especially the production of its journal literature, is a historically and culturally contingent practice that produces and disciplines the discursive resources required to generate a narrative in which laboratory phenomena are stripped of all locality, situatedness, and contingency. Through documentary practices, they are transformed into phenomena of an objective world. The formal literature is therefore an important part of the cultural and historical monument of modern, Northern/Western science. What we recognize as science would not be what it is without scientific documentation. But our kind of written science is no more necessary to the development of stable and enduring knowledge of the natural world than the classical form of music – another phenomenon thoroughly intertwined with writing – is essential to the existence of music per se.

This chapter begins with the difficulties faced by science studies that view all of scientific labour primarily as the production of statements, rather than viewing statements as just one kind of the 'bits and pieces' constituting the 'motley' of the sciences. The work of Latour and Woolgar (1986) and Ravetz (1971) is important in this regard because they explicitly address the role of writing in scientific practices. Knorr-Cetina's (1981) careful investigation of how a typical contemporary scientific article is produced, from laboratory practice through successive drafts to final product, shows the complex and thoroughly non-representational relationships that exist between statement production and laboratory work. In developing her conclusion, that we must abandon the idea of the article as a faithful description of laboratory practice, she gives us important conceptual tools for analysing the role of scientific writing. Her insistence on how the contemporary article constructs a particular kind of space – an information space – as the site of scientific facts invites comparisons with historical studies, especially Shapin and Schaffer's (1985) analysis of Restoration England's literary technology of science as an objectifying resource.

Bazerman (1988a), Rouse (1996), and Fleck (1979) each raise the issue of the documentary practices of science in ways that challenge epistemological reductions of the significance of documents for knowledge production to the communication of information. Bazerman's inquiry into the reading practices of physicists shows that the formal literature provides resources to situate laboratory labour in a field of significance understood as a narrative of collaborative practices in which one's own work plays a specific role. His study has affinities with Rouse's investigation of the narrative structure of the practical understanding manifested in scientific labour. Rouse's analysis also provides conceptual tools to understand the relationship between writing and non-documentary practices of scientific work. Fleck's typology of the stratification of scientific literature permits more precise questions to be posed about the distribution of objectifying resources over document types. Taken together, the studies canvassed here provide a set of analytical resources that focus attention on constitutive rather than communicative effects of scientific writing.

Science as Writing

Studies of information needs and uses of scientists have shown that analyses of the journal article must acknowledge that it is not used to derive new results at the research front. We have seen that a common response to this finding is to recuperate the article's significance as a token exchanged for reward and status. According to this view, the paper has a dream-like structure, in which its talk of protein synthesis or the isolation of free quarks becomes the disguised expression of its true, latent content: a claim for priority in science's hierarchical social system of professional status. As a research resource, the journal article is banished to the exoteric zones of scientific labour. Any plausible account of its importance, however, must find some way to take into account the stories told in the formal literature. One way to establish a very strong connection between science and its formal literature is to construe the journal article as the primary product of scientific labour. We have already encountered the IC-model version of this view, in which the article is the material vehicle of science's immaterial, primary product: information. Yet there are non-IC versions which, although they avoid the idealizations of IC models, are vulnerable to conclusions of use studies. To show this, I consider briefly two analyses from studies of scientific practices which interpret articles as the primary products of science.

Modifying Statements, Building Black Boxes

In *Science in Action* (1987) Latour considers 'the anatomy of the most important and the least studied of all rhetorical vehicles: the scientific article' (31). He sees the article not as a vehicle for communicating information but as a resource for stabilizing statements. The centrepiece of his analysis is the notion of a fact as an unmodified sentence. If a sentence situated in the authorized documentary circuits of the sciences is unqualified, unchallenged, and cited again and again without question, it becomes a fact or, as he puts it, a 'black box' analogous to technological artifacts upon whose routine use scientists come to depend. Although his focus on the production and status of 'facts' indicates a residual fidelity to epistemological language, whatever traditional epistemological concerns remain relevant for Latour are radically materialized. Writing a journal article is a labour process because facts are constructed, and their construction requires a technology of writing. Much of the work of stabilizing statements is performed in the scientific literature itself, and can be revealed by tracing the various rhetorical techniques aimed at stripping rival statements of all modalities other than the assertoric. Anyone who wishes to dispute a factual claim or reopen a 'black box' in order to critically examine the processes of its production must be forced to exert an unacceptable amount of labour. Latour is fond of military analogies: bestowing the status of fact by marshalling other facts in support is like mobilizing the troops. References become crucial: 'The presence or absence of references, quotations and footnotes is so much a sign that a document is serious or not that you can transform a fact into fiction or a fiction into a fact just by adding or subtracting references ... A paper that does not have references is like a child without an escort walking at night in a big city it does not know: isolated, lost, anything may happen to it' (33). Stabilizing a statement such that it is used routinely and without question by others therefore demands considerable effort and skill in manipulating the archive of scientific papers.

In Latour's view the strategies governing the stabilization of statements in the journal literature are basically the same as those of experimental laboratory work. In *Laboratory Life* (1986), he and Wollgar argue that laboratories are machines for constructing statements: 'A laboratory is constantly performing operations on statements; adding modalities, citing, enhancing, diminishing, borrowing, and proposing new combinations ... in situations where a statement is quickly borrowed, used and reused, there quickly comes a stage where it is no longer contested. Amid the general Brownian agitation, a fact has then been constituted ... To this end rats had been bled and beheaded, frogs had been flayed, chemicals consumed, time spent, careers had been made or broken, and inscription devices had been manufactured and accumulated within the laboratory. This, indeed, was the very raison d'être of the laboratory' (86–8). Once a statement is stabilized in the laboratory, it is inscribed in the literature, where it is vulnerable to further destabilization, but only at the high cost of deploying another laboratory to stabilize its rivals. Because Latour sees the construction of stable statements as the end of all scientific work, networks of inscriptions constitute for him the totality of the scientific enterprise. Such networks include a wide array of actors, technologies, social disciplines, and institutions engaged in strategies and practices of statement stabilization. The work performed in the scientific literature provides a template for scientific work in general. For Latour, documents and documentary circuits are everywhere; he follows scientists and engineers until he finally tracks them down in 'centres of calculation,' which turn out to be document processing sites (1987, chapter 6).

Latour therefore places the formal literature at the centre of the action. Yet there is a huge distance between his view and those that privilege information and communication. Contrary to epistemological models, sentences have no epistemic status in and of themselves. *'By itself,'* he writes, *'a given sentence is neither a fact nor a fiction; it is made so by others, later on* ... scientists, engineers and politicians constantly offer us rich material by transforming one another's statements in the direction of fact or of fiction' (25; emphasis in original). Statements are always up for grabs, at least in principle, and when they do stabilize, they do so only because challenging them has been made too difficult. A single scientific statement cannot by itself communicate information, because, in and of itself, it has no factual content. Its 'facticity' is a function of its degree of stability, and that is determined by what other actors in the network do to it. Whether or not a statement becomes informing is therefore dependent upon the temporally extended actions of a multitude of agents. 'Communication' and 'information' become *effects* of the shifting stablilities of statements.

The great advantage of Latour's analysis is that it reveals how the statements of science's formal literature operate in the laboratory: some are deployed to stabilize new statements and some to destabilize others. For him, science's formal literature is crucial because it is the site where facts emerge. He sees science's literary technology not only as the pinnacle of the process of fact construction, but also as its paradigm, exemplar, and most illuminating metaphor. Yet in spite of all the attention he pays to the massive labour required to install the apparatus required to produce facts, his view of science as a network of statements has an epistemological flavour. Although there is little room in his picture of scientific labour for the 'information seeking' featured so prominently in IC models, non-discursive scientific labour and its products are subordinated to the production of statements. Latour has done much to show how documentary practices intersect with laboratory labour, but at the expense of exhausting the purpose of the non-discursive scientific practices that other studies have shown to have a life of their own by reduction to fact generation through statement stabilization.[2]

Like Latour and Woolgar, Ravetz (1971) does not privilege the communication of information in scientific practices, nor does he view the extensions of scientific work from laboratory to laboratory in such terms. These authors reject traditional, epistemological accounts of scientific rationality as idealizations of scientific method. They also reject the idea of a theoretical representation of the world as the goal of scientific activity, recognizing that scientific labour is localized, opportunistic, and contingent. 'Information' finds no resting place in enduring facts in their accounts, because 'facts stay alive only as long as they are useful in new contexts, and if they cannot be put into a convenient form for their limited functions, they are soon discarded and forgotten' (209). Thus for Ravetz, as for Latour and Woolgar, the universalist tendencies of IC models are explicitly disavowed. Yet they agree that scientific activity aims at statement production. Ravetz holds that the primary product of scientific work – 'a special sort of craft work' (116), 'hard, complex and sophisticated' (133) – is a statement. He sees such work as a response to a problem, and he defines a scientific problem itself as a statement.[3] A consequence of what might be called the 'textual' view of Latour, Ravetz, and Latour and Woolgar is the centrality of the journal literature to scientific practices.[4] However, the rich descriptions of the varieties of scientific labour provided by many studies of practices show that their reduction to writing is implausible. Scientific writing is but one of the kinds of labour performed in the production of scientific knowledge. The 'textual' view of science tries to reconcile the sphere of writing with laboratory life but at the expense of an implausible reductionism that oversimplifies the relationships between writing and other activities scientists pursue. In this respect it has similarities with the simplifying tendencies of epistemological models of science.

Ravetz distills the essence of science – its statements – into a specific documentary form, the 'research report,' which he sees as 'the product of a completed scientific investigation' (182). If statements are to become facts, they must pass certain tests. The first is the certification of journal article manuscripts. Without institutional certification, Ravetz writes, the research report has no value: it 'must be checked by an impartial assessor before it is released to the community. This is commonly done by a referee for a scientific journal; only after the referee has certified the problem as being of sufficient value, and the work as having been performed to the ruling criteria of adequacy, can it appear in the literature. A research report which is distributed without this certificate is, or traditionally has been, considered as nearly worthless' (182–3). The second test is communal: the paper must be interesting to at least some other scientists. In language reminiscent of Latour's metaphor of the child lost in the city, Ravetz writes: 'the automatic, communal test of value yields a negative result when a particular paper happens to be of interest to absolutely no one. As a candidate for the status of scientific knowledge, it has then been killed ... as a contribution to the advancement of knowledge it has died, precisely because it is unwept, unhonoured, and unsung' (184). A legitimate scientific paper functions as a resource for further work: its statements must be among those used to produce further statements, which in turn are submitted to the cycle of certification, communal interest, and resource use.

The reduction of scientific resources to certified statements presents a difficulty for Ravetz's account. Since the statements he takes to be the primary products of scientific activity must be produced from other certified statements, those not so produced are generated from worthless statements, thus remaining worthless themselves. But, as use studies show, the process of certification takes too much time. Scientific labour does not wait for statements to be certified. Thus the reduction of scientific labour to the production of certified statements from previously certified statements runs afoul of use studies. The accounts provided by Latour, Ravetz, and Latour and Woolgar are very illuminating as descriptions of the work involved in generating new literature from old. But the relationships between the labour of writing and the non-discursive labour processes of science become highly problematic, and doubly so when scientific research practices are held to require certified statements. Non-IC versions of the journal article as the primary product of scientific labour fare no better than IC versions in the face of use studies.

Although I have argued that scientific writing is properly construed as the production of statements in the Foucauldian sense, scientific labour as a whole consists in more than writing. Statements are among the matériel of science. The problem is to show how statements – science's documentary matériel – link up with the other elements involved in scientific practices. It is implausible that the research-front activities enumerated by Latour, Ravetz, and Latour and Woolgar consist in producing a product – a scientific text – that few read, and, so it follows from their analysis, that is taken seriously primarily to write a new – yet already old – scientific text. This reductionism depicts science as kind of glass bead game in which massive resources are pumped into an activity whose product is continually recycled in the production of a similar product, which is in turn recycled, and so on ad infinitum. Yet other studies of scientific practices show that there are products of the laboratory beyond texts: the non-literary black boxes, the disciplined routines, the epistemic alignments, the social networks, the machines, the technologies – all the things that Latour himself insists upon in his later work but subordinates to statement production.[5] The problem posed by such studies for scientific documentation is what role documents play in the assemblages linking these disparate kinds of elements, and how the cultural imperative to produce the formal literature of science configures both discursive and non-discursive laboratory labour.

Conversions and Perversions

Content analyses of journal articles show that they do not represent the process of scientific discovery, but present after-the-fact proof, omitting false leads, unsuccessful efforts, and the factors determining both the choice of problem and experimental method. Not only do they typically fail to provide enough information for the replication of successful experiments, but the very possibility of replication has been called into question.[6] Such analyses pose serious problems for epistemological accounts of journal articles. Bazerman puts the challenge most pointedly when he asks (1983, 158): 'If a scientific paper is not a complete account of a scientist's observations and doings, nor a tightly argued deductive proof of claims, nor an unproblematic conveyor of claims to be objectively evaluated fairly and promptly by a professional audience, what indeed is the scientific paper communicating, and to whom?'[7]

One of the most sustained attempts to come to grips with the details of scientific writing is Knorr-Cetina's (1981) case study of the production of journal articles. Her analysis of scientific labour, noted in the previous chapter, emphasizes the opportunistic, contextual, and contingent reasoning that characterizes the selection of highly localized laboratory resources. In her case study she shows that the journal article also exhibits a complex relation to contingency. Although its production is marked by the same opportunism, contextualization, and contingencies of other laboratory products, its content explicitly denies and disavows not only the contingencies of its own production, but also those of the laboratory work which it purports to represent and of which it purports to offer a faithful report. 'In the laboratory,' she writes, 'scientific reasoning displays its concerns in savage purity. But in their papers, the savage reasoners of the laboratory seem to change their faith. The reasoning of the paper, one imagines, is faithful to the scriptures (the authoritative writings) of an area, rather than to the concerns from which it originates' (94).

Three key ideas arising from Knorr-Cetina's investigations are especially important for understanding the role of the journal article in scientific practices. The first is scientific writing's 'double-threaded framework' of decontextualization and recontextualization. The paper not only occludes the local circumstances in which resource selections are made, thus decontextualizing laboratory reasoning by stripping it of its opportunism and locality, but constructs a new context in which laboratory work and its results are recontextualized in an alternate set of meanings. The second is the rhetoric of hesitation, doubt, and reluctance to draw conclusions, a literary form whose origins Shapin and Schaffer (1985) locate in Restoration England. The third is the interpretation of the literary technology of science as an objectifying resource that helps construct scientific objectivity as a discursive product of techniques and technologies of writing.

Decontextualizations and Recontextualizations

Scientific writing's 'double-threaded framework' may be traced through Knorr-Cetina's investigation of how the three standard sections of the journal article – its *Introduction, Methods and Materials*, and *Results and Discussion* – are constructed. A paper's *Introduction* provides the framework which establishes its meaning and significance. But it does so not by situating methods and results in the context of the practical reasoning that governed laboratory decisions and resource selections.

Whatever relevance is constructed in the *Introduction* for the rest of the paper will not permit a reconstruction of the problem that led the laboratory scientists to their methodological selections or of their practical reasoning in pursuit of a solution. The *Introduction*, writes Knorr-Cetina, 'is the only place we can turn for an answer to the question "why" in respect to methodological selection. Yet clearly the reasoning of the *Introduction* does not contain the answers to such questions' (1981, 118). Instead, it provides an alternative framework of relevance. It generates new meanings and new significance for methods and results by telling a *simpler story*, one conforming more closely to the canonical logic of scientific rationality in which complexities of chance and contingency yield to uniformities of idealized formal method. Rather than emerging from a complex and tangled, heterogeneous and transscientific field of practical reasoning, the work of the laboratory is recontextualized in an *information space*: an abstract realm inhabited by problems and solutions. A Ravetzian drama of scientific activity where problem statements meet solution statements is indeed staged, but in the scientific paper, not in the laboratory. Knorr-Cetina sees the journal article as the script of this drama: 'The impression of a problem-pushed solution which has been researched, rather than encountered by chance, is created in the text through the hierarchical organisation of arguments through which the solution appears *derived* rather than original' (101).

The *Introduction*'s recontextualization is therefore a fiction. The problem context it presents is not the practical context of the laboratory. The problem is recontextualized as a *type*, one that *could conceivably* have motivated not only this but other laboratory work. As Knorr-Cetina puts it: 'To be accepted for publication, the resource reasoning must provide a *plausible script* ... and not a scenario which has been or will be realised' (112). The paper is therefore not a *representation of* but a *move in* a field of resource relationships and resource conversions. It is a phenomenon of institutionalized discourse about these processes, not a phenomenon of the processes themselves. If the paper is to become a resource for further work, it will be because others have found in it a plausible script for documenting their own work, not because it is an accurate record of the problem the laboratory scientists faced. The only social action with a correlate in the paper is located in an anticipated future which the paper attempts to bring about, one which depends upon 'the practical response of those the paper manages to interest' (112).

The 'double-threaded framework' of the paper's *Introduction* is absent from its *Methods and Materials* section. Here, only decontextualization is at work. The absence of context is not surprising. To present the reasons for laboratory resource selections would run afoul of the *Introduction*'s fictional framework of relevance. Knorr-Cetina points out that 'the typification of the paper's version of *Method* converts the painfully constructed "way" (or method) of the laboratory into a *natural consequence* of the work's overall purpose and the reasoning contained in the *Introduction*' (118). The smooth transition between the two sections of the paper is a literary artifact. *Methods and Materials*, she writes, 'resembles, more than anything else, the recitation of a formula. We find, not laboratory tasks, but a laconic checklist of steps taken ... a catalogue of sequential manipulations stripped of both context and rationale.' The checklist appears to flow naturally from the work's fictional purpose only because methodological options are presented as if they were not *selected* at all: 'method is presented as a flow-chart of selections disguised as *non-selections*, for lack of relevant contextualisation' (115). Moreover, the checklist itself is highly selective. Because it fails to include 'all the information relevant to the technical success of an outcome, or even relevant to an informed evaluation of the results' (129), it cannot be construed as a representation, or summary description, of methods actually pursued in the laboratory.[8]

The selectivity exhibited in *Methods* might be thought to exclude those elements quite properly excluded from any reasonable description, which Knorr-Cetina describes as one 'which refers only to those actions, events, and phenomena which are immediately relevant for obtaining the paper's technical results,' or 'which contains only that information which cannot be reasonably presupposed, and is not entailed or a likely consequence of what is said' (127). Much tacit knowledge and many activities performed in the laboratory are simply not relevant to a description of an experiment. A faithful description of scientific labour, like any description of a complex activity, cannot be expected to contain every detail. However, she shows that the paper excludes not only much practical and local know-how that bears directly on obtaining the paper's reported results, but also what is required to reconstruct 'the recipe sequence of steps in the paper into that of feasible doings' and 'the routines for diagnosing and coping with many unspecified problems' (128). The failure to include such know-how indicates the irrelevance of the local contingencies of practical, laboratory reasoning to the discursive project of telling a story consistent with idealized scientific method. Nor does this failure present a special problem for scientists, since the replication of experimental results that might depend upon an experimental report in the form of a summary description or recipe is not only rarely, if ever, performed, as

Hacking has pointed out, but, as Collins (1992) has suggested, may even be impossible in principle.[9]

Thus the decontextualizing strategies of *Methods* stage the second act of the scientific paper's drama of scientific rationality. They ensure a seamless connection between the *Introduction*'s abstract problem space and the steps 'reported' in *Methods*. 'The point here is not that the *whole* story is missing, but that *any* (technical) argument which accounts for a (final) choice, as well as any problematisation of alternative possibilities, is rigidly avoided in the section of the paper which purports to be a report on laboratory procedure. Compared with the relevant work in the laboratory, where the *making* of selections dominates the scene, the paper offers a curiously purged *residual* description, constituted more by what is *not* at stake in the research ... than what is' (115). *Methods* stages the activities of the laboratory in an absence of any specific context, thereby clearing a space for the entrance of the impersonal, methodical subject of universal, scientific reason.

The third and final act is staged in *Results and Discussion*. In the practical contexts of laboratory work, results and method are interdependent. Yet in this final section of the article, the dependence of results on methods is explicitly disavowed. In the paper, laboratory results achieve an independence and singularity enjoyed by epistemological conceptions of *information*: thoroughly decontextualized, they exist only in relationships to each other. Documented results therefore inhabit the same information space as the paper's problem, 'a space far distant from the laboratory.'[10] Their disassociation from the methodological selections of the laboratory is reinforced by the reasoning of *Results*, which typically forswears drawing conclusions at all, even though a future may be envisioned in which conclusions might be drawn. 'The pattern is well suited to the stereotyped image of science as presenting the "facts" which others may use in making decisions' (123).[11] In the final act of the article's drama of scientific reason 'facts' emerge, immobilized on a stage of abstract and decontextualized scientific rationality, framed by the *Introduction*'s fictional problem. It is a drama of epistemic content performed in writing.

Modest Witnesses

The reluctance to draw conclusions in *Results* so that facts can 'speak for themselves' is one aspect of the journal article's characteristic rhetorical style. Other aspects are traced through Knorr-Cetina's study of the changes between the first and final – in her case, the sixteenth – draft of the article manuscript. She identifies three major strategies of revision. These consist in '*deleting* particular statements made in the original version, *changing the modality* of certain assertions, and *reshuffling* the original statements.' The general strategy is to soften dramatic impact. The deleted statements are either those which had increased assertive force by reinforcing previous points, or which were considered assailable, hence 'weak' or 'dangerous.' Assertive force is also decreased through changing modalities 'from the necessary to the possible, and generally from the strongly asserted to the more weakly asserted ... The scientists either hold back their claims, or couch them in terms which denote hesitation or doubt.'[12] The strategy of reshuffling also diminishes impact, 'resulting in a loss of clarity and straightforwardness.' For example, the section on the work's purpose is revised such that it is not stated directly, but proceeds in the final version 'by a spiral-like *hedging in* on the purpose it gives to the study' (102). Also, the first draft's clear assertion of the importance of the results as a significant discovery with specific applications, albeit not in the actual but in an improved, possible world 'has also disappeared: the mission of the published paper is no more than a comparative analysis and evaluation' (103).

The goal of this literary style is to *maximize believability* by occluding or at least minimizing any subjective, controversial, and apodictic assertions, thereby softening strong claims in favour of those weakened by submission to reasonable doubt. Insofar as any narrative imagines its author, journal writing projects a modest, rigorously empirical, non-dogmatic subject content to witness the workings of an objective world with a prudent reluctance to pass judgment upon it. This subject's investigative tools and procedures are fallible, and only appropriately qualified statements are presented for scrutiny by equally reasonable scientific peers. Because the process of softening the paper's rhetoric results from its co-production, which incorporates the revisions of readers, the author becomes a communal, corporate subject. The scientific paper therefore enacts the workings of a modest, objective, and collective mind confronting an objective world with detached, scientific discernment.

The rhetorical strategies appropriate to the 'modesty' on display in Knorr-Cetina's study have a venerable ancestry in the history of science. They may be traced to the 'literary technology' devised by Robert Boyle, who created and disciplined a particular form of scientific writing that founded the modern journal article, a form characterized by Shapin and Schaffer (1985) as 'the modesty of experimental narrative.' Borrowing a term from Donna Haraway, we may call it the style of the 'modest witness.'[13] In Restoration England, Shapin and Schaffer write, 'the literary display of a certain sort of morality was a technique in the making of matters of fact. A man whose narratives could be credited as mirrors of reality was a *modest man*; his reports ought to make that modesty visible. 'In the early period of the Royal Society, to strike a posture of modesty through scientific writing consists, first, in eschewing grand natural-philosophical systems for the piecemeal work of scientific journeymen who are satisfied to write experimental reports with limited goals. The natural philosophers 'who wrote entire systems were identified as "confident" individuals, whose ambition extended beyond what was proper or possible. By contrast, those who wrote experimental essays were "sober and modest men," "diligent and judicious" philosophers, who did not "assert more than they can prove"' (65). And proof, in experimental matters, required that all traces of personal style be purged from the writing, to let facts speak for themselves. Such a style 'served to display ... the philosopher's dedication to community service rather than to his personal reputation (66).[14]

Exploring similarities between contemporary scientific writing and Boyle's literary technology of modest witnessing is not meant to suggest that the experimental report has not changed in either purpose or form since the early half of the seventeenth century. Indeed, Peter Dear (1995) argues that Boyle's work and that of the early period of the Royal Society represented a short-lived and geographically isolated form of 'experimental philosophy,' one subsequently reformed by Newton's 'physico-mathematics,' which provided a philosophically richer way of generating universal science from Boyle's singular, experimental events and his Baconian focus on facticity alone.[15] The evolution of the scientific report is an interesting question in its own right, but it is not pursued here.[16] Yet it is still the case that grand schemes, usually published as books, have a lower epistemic status than journal articles. Furthermore, the decontextualization of *Methods* and *Results* takes the stylistic form of a flat, unadorned recitation of events. Perhaps the most striking similarity between Boyle's and contemporary ways of presenting facts with sufficient stability, as he puts it, to 'make their own way,' is that each inscribes a discursive separation of matters of fact and the more speculative statements in which they are embedded or which refer to them. Shapin and Schaffer note that in Boyle's literary technology, there 'were to be appropriate moral postures, and appropriate modes of speech, for epistemological items on either side of the important boundary that separated matters of fact from the locutions used to account for them: theories, hypotheses, speculations, and the like' (66–7). For matters of fact, 'a confident mode was not only permissible but necessary.' But here is Boyle's advice to his nephew on the proper style for venturing what he calls *opinions* in the experimental report: 'in almost every one of the following essays I ... speak so doubtingly, and use so often, *perhaps, it seems, it is not improbable,* and other such expressions, as argue a diffidence of the truth of the opinions I incline to, and that I should be so shy of laying down principles, and sometimes of so much as venturing at explications' (67). Boyle's nephew appears in this passage as the spiritual ancestor of those readers Knorr-Cetina found to be responsible for 'softening' the journal article's drafts. And the distinction between the confident style for matters of fact and a hesitant style for more speculative and interpretive assertions is also mirrored in Knorr-Cetina's study, in the contrast between the plain speaking of *Methods* together with the more interpretive problem setting of the *Introduction* and the reluctance to draw conclusions in *Results*. Whatever other changes the experimental report has undergone in the last three and a half centuries, the stubborn durability of the literary rhetoric of 'modesty' in scientific writing testifies to its historical ancestry.

Objectifying Resources

Commenting on Robert Boyle's documentary innovations, Shapin and Schaffer remark that 'the objectivity of the experimental matter of fact was an artifact of certain forms of discourse' (1985, 77–8). Because Boyle's experimental reports were designed to present matters of fact as *given* items, they call his literary technology an *objectifying resource.* 'If the obligation to assent to items of knowledge was not to come from human coercion,' they ask, 'where did it come from? It was to be nature, not man, that enforced assent. One was to believe, and say one believed, in matters of fact because they reflected the structure of natural reality ... *Yet the transposition onto nature of experimental knowledge depended upon the routinization of these technologies and conventions*' (79; emphasis added). Knorr-Cetina's studies show that formal scientific writing still functions as an objectifying resource, through decontextualizations and recontextualizations, eliding particularity, and reinscribing scientific work in an abstract information space of facts, problems, and solutions.

The scientific paper works as an objectifying resource not because it represents the real world of the laboratory; it instead discursively constructs an alternate world. It is not the objectivity of accurate reportage but of literary construction. It draws its resources from two sets of inscriptions: those of the laboratory and those of the scientific literature. The section generally written first, *Results and Discussion*, originates from the former – in Knorr-Cetina's case study, from 'the measurement data and laboratory protocols, whose numbers, graphs, and photographs we find in the published paper in a cleaned-up, composed, and edited version prepared by the "art-shop" of the institute' (1981, 130). *Methods and Materials* also draws upon laboratory inscriptions; in the case study, it was derived from 'the flow chart of experimental steps prepared by the scientists for the technicians who were to run the ... tests' (129–30). The formal literature is the source of the *Introduction*'s fictional context of relevance; Knorr-Cetina says it 'derives from the scriptures of the area' (130). The result is a transformation of laboratory labour that she calls a 'conversion of reason.' It is a conversion *from* the localized, contingent, opportunistic, highly situated, analogical, and practical reasoning governing laboratory resource selection *to* the abstract, decontextualized, and objective information space of scientific papers. The recontextualizations of laboratory decisions and resource selections present facts that 'make their own way' on a stage whose objectivity depends upon the discursive erasure of all traces of human intervention.[17]

Given the disequivalence between laboratory reason and its discursive construction in the scientific paper, its links to laboratory work cannot be explained in cognitive or representational terms. 'The scientists who write a manuscript,' observes Knorr-Cetina, 'do not recall the research process and then proceed to summarise their recollections' (130). The paper is a specific kind of *discursive product*, different from the laboratory's material set-ups, but no less an outcome of scientific labour. Formal writing creates an information space of problems, steps taken, projected futures, and specific results embedded in networks of similarity and difference. Its rhetoric performs a dual virtual witnessing, not of laboratory reasoning and practices, but of isolated 'facts': not only are the witnesses not present at the scene, but the scene itself is a discursive construct. The scientific paper's virtual witnessing consists in making present *a simulacrum as witnessed*, through which it gains the status of an *object*. Insofar as knowledge of an experiment comes to depend upon its reconstruction in the journal article, the real laboratory situation is forever erased: 'Except in the memory of those who were present during the process, it is an irreversible transition' (130). In modern science, objectivity is written in terms of abstractions; decontextualized facts floating in information space are its signature.[18] Creating such a space is not the same as building vehicles to communicate information.

The transformation of practical into cognitive labour is not only a conversion but also a perversion. 'Compared with the work observed in the laboratory,' writes Knorr-Cetina, 'the written paper is, as we have seen, a first complete perversion,' just as the 'transfiguration in a fairy tale of a wizard into a mouse is at the same time a disfiguration of the wizard' (132). This 'perversion' is not a failing. Instead, it exemplifies a general feature of resource conversion. Scientific work aims at launching resources for other research contexts. Rouse's concept of epistemic alignments, Latour's insistence on the strength of associations, Pickering's 'mangle of practice,' and Rheinberger's notion of 'epistemic things' all make this point. Knorr-Cetina views scientific practice as the labour of constructing objects which are 'continuously and asymmetrically ... *reconstructed* from a preceding object, while their *equivalence* to or *difference* from these preceding objects is at the same time *negotiated*.' She points out that this kind of labour operates not in a context of transmissions and flows, but of transformations. 'In economic terms,' she writes, 'what is postulated here is an *economy of change* rather than of exchange, a process in which equivalence is superimposed upon disequivalence, and in which disequivalence means conversion as well as perversion' (131–2). Smooth communication of a 'noble substance' that maintains its integrity no matter what its vehicle has no place in such an economy, except as the product of documentary work.

In Knorr-Cetina's analysis, scientific documentation plays a key role, not only as an outcome of laboratory work, but also as a resource used to configure further work at the research front. 'Writing itself,' she observes, 'is an apt medium for ... perversion.' Presumably, its aptness consists in the relative ease with which discursive resources can be manipulated to effect the paper's 'conversion of reason' from laboratory context to abstract information space. The 'tinkering' performed to 'make things work' is, presumably, easier when the thing made to work is a paper rather than a complex laboratory set-up. But all scientifically meaningful laboratory objects 'undergo a recontextualisation and reconstruction similar to what we found in the writing of the paper.' In language reminiscent of Latour and Ravetz, Knorr-Cetina notes that laboratory resources that are not converted 'into the ongoing enterprises of other social agents will be neglected and ignored.' Such conversion, she points out, is actively sought in the laboratory. The significance of the scientific paper is that it provides a 'script of resource conversion' (132). Its elision of particularity, its decontextualizations and recontextualizations, are what permit the paper to present a plausible scenario for the conversion of its itemized resources – decontextualized actions, fictional frameworks of relevance, and isolated results – for others not only to document their own work, but to configure it such that it can be so documented. The scientific paper, far from communicating information, presents an abstract, *possible world* that laboratory work must be made to inhabit.

Enacting Narratives

The concept of the scientific paper as a script for resource conversions actively sought in the laboratory moves analysis of scientific documentation beyond the production processes of science's literary technology and installs it at research-front work. Many studies tell us much about strategies and tactics of writing journal articles. Documentary production processes, however, tell only half the story. The previous chapter asked how journal articles figure among the 'bits and pieces' of the heterogeneous resources involved in the production of scientific phenomena. Exclusive attention to how science's formal literature is written tends to confine analysis of the resource role of scientific documents to production of other documents. In making this point I do not wish to deny the salience, among scientific activities, of producing new literature from old.[19] Because writing is an important part of scientific culture, investigations of production processes of scientific writing do move analysis beyond the marginalizations of the resource value of science's formal literature found in epistemological models. Moreover, the studies indicated here have shown that deployments of previous literature in scientific writing do not adhere to standards of rationality featured in epistemological models because selections of documentary resources to produce new documents exhibit the same situated, practical reasoning that governs laboratory decisions. And if we accept Latour and Woolgar's assumption that all scientific activity consists in the production of statements, then writing articles figures in research-front activity even if they are used only to write other articles, simply because there is no clear line of demarcation between the work of stabilizing statements in the laboratory and in the literature. Most studies of practices do not share Latour and Woolgar's textual reductionism, but show instead that scientists use a wide variety of matériel to produce a correspondingly wide range of scientific phenomena. Moreover, if scientists produce more than just statements, and if the formal literature configures non-documentary production processes, then it is directly relevant to research-front work.

Knorr-Cetina's 'script of resource conversion' is a move in this direction. It shows that the relationships of influence and effect between laboratory and documentary labour are not unidirectional: it is not the case that inscriptions move only from the laboratory to the article. As a script of resource conversion, the article embodies statements that provide a framework of reasoning about the configuration and conduct of laboratory work. It becomes a social object, coordinating the work of scientific and transscientific fields.

Rouse's concept of the narrative structure of scientific work may be read as an elaboration of Knorr-Cetina's insight. Although not going so far as to echo views of scientific writing as an agonistic field (Latour and Woolgar 1986) or even guerrilla warfare (Knorr-Cetina 1981, 126), Rouse recognizes that science's documentary practices involve situating one's work in an ongoing narrative of scientific work in general. He writes: 'Scientific experiments are designed, and papers written, in response to the specifically anticipated concerns of other scientists rather than some context-free standard of rationality' (1987, 121). If an ongoing narrative configures scientific labour, scientific writing configures such labour to the degree that it determines the narrative. A study by Bazerman of physicists' use of the journal literature provides a valuable introduction to Rouse's analysis of the narrative structure of scientific practices.

Reading Physics

Bazerman's (1988a) small study of the reading patterns of seven physicists moves a step beyond epistemological marginalizations of the role of the journal article. By challenging the idea that research-front work is an information-gathering activity, his work suggests an alternate role for journal articles, one that situates them much closer to the action. He found that the reading of the physicists he studied had a dual relationship to their work. First, their selections of articles, those they retained and what they got from them, were based upon their own research projects, both current and anticipated. Their research was a resource for attributing significance to articles, not a search for fresh epistemic content for the further development of a theoretical representation of the natural world. Contrary to IC models, Bazerman's study shows that the physicists' research does not generate criteria to reject the journal literature for their current work only to use it instead for non research-related tasks such as general awareness, keeping current, and the like.

Second, the significance of the physicists' research derives from their grasp of the field and their place in it, which Bazerman calls a 'personal map or schema of the field.' By embodying 'the physicist's personal perception of the forward motion of the discipline of which the researcher considers himself or herself a part' (242), this schema provides criteria for the selection of articles, thus situating one's own work in relation to that of others. Especially important were articles that might spur a change of research direction: 'Articles, in their challenge to existing statements, foment new work. Plausible new methods, evidence, claims, and interpretations change the landscape against which the researcher plans and realizes research purposes' (249–50). Yet the physicists' personal schemas were themselves derived from the literature: 'over the long term the body of claims from the corporate literature that are integrated into the individual's schema will close off certain problems and methods and open up others. A changing picture of nature and the dynamics of investigation, all garnered from reading, will modify research purposes ... the literature is ... understood, criticized, and evaluated against an image gleaned from the literature itself rather than against nature itself' (250). The personal schema's environment turns out to be thoroughly documentary, literary, and discursive.

Bazerman's study is useful for understanding the narrative structure of scientific work in spite of the mentalistic implications of its talk of schemas, personal maps, pictures, and perceptions. Such implications are most pronounced in references to scientists as information processors: 'The reader will process information that has significance for the existing schema and will view that information from the perspective of the schema' (243). The implicit mentalist and epistemological assumptions at work here interpret the activities of attributing significance as taking place in a hidden, mental realm. Situating one's own work in the context of collective scientific endeavours involves an intermediary step of developing and altering a schema described as a 'personal perception.' The problem with explanations of this sort is that it is never clear what a mental state is supposed to be. But Bazerman's observations hold much promise if restated in a material register in which 'concrete work,' as Latour puts it, replaces mental activities.

For Bazerman, scientists aim at the production of a literature, not at the production of individual statements. Statements enunciate facts, but a literature tells a tale. Bazerman therefore makes a distinction between statement production and literature production. The latter leads to the importance of narrative in scientific work because it involves the joint effort of constructing what Bazerman calls 'a view of nature' (250). But the process of constructing a literature, thereby continuing the tale told by scientific research, must first move through the laboratory. Literature production is not a direct transformation of one set of literary inscriptions into another. In contrast to models of science as information processing, which relegate the journal article to the periphery of research work, and in spite of his references to the scientist as an information processor, Bazerman views the journal literature as a vital resource for establishing the significance and intelligibility of current scientific research. He interprets reading physics as integral to positioning research work in an ongoing narrative of science. In this way, the journal literature configures laboratory work itself. Its value is not exhausted in the production of a 'personal schema,' but depends upon documentary practices of deploying inscriptions from the literature, not only to stabilize statements, but also to configure material laboratory set-ups such that they can be narrated as belonging to ongoing, collective scientific work.

By centring the journal literature in this way, Bazerman's work raises issues for further study. Scientific documentation is not the only discursive resource for the articulation of the significance and intelligibility of scientific results. Tracing the distinctions between different kinds of resources made available by the various institutions with a stake in scientific results points in one research direction; related issues concern the relative importance of such resources, their location on the cultural terrain, and how they configure what gets said in the name of science both within and beyond laboratory walls. Bazerman's small study of seven physicists raises research questions: Is it possible to identify and describe in some detail the various historical and cultural forms of intelligibility and significance fabricated from the different kinds of discursive resources specific to various practices of knowledge production? And if so, how do these forms of intelligibility intersect with cultural practices beyond institutionally legitimated research science?[20] These are questions about the social roles of various forms of scientific documentation, not about the communication of information.

A thorough analysis of science's literary technology therefore encompasses both its production processes and its resource uses. It situates the journal article not only as the end-product of a highly specific form of writing, but also shows how and to what extent the literature configures and is intertwined with the many other practices that together constitute the labour of scientific research. In Pickering's terms, such an analysis reveals the role of the documents in the mangle of practice. Stripped of its mentalistic language, Bazerman's study points in this direction by its connection between the tale told in science's formal literature and the conduct of research work.

The Narrativity of Scientific Practices

Rouse's analysis of the importance of narrative to scientific work permits a deeper exploration of these issues without leading into epistemological and mentalistic blind alleys. Like other writers on scientific practices, he insists on the local, situated, and highly contingent character of scientific knowledge. But if scientific phenomena are local, how can they be significant to collective endeavours? Rouse formulates the question as follows: 'Science is practiced in scattered locations. Its practitioners come from varied cultural and scientific backgrounds, work in different material settings, and often work toward disparate proximate aims. Yet theirs is nevertheless a shared enterprise: research aims to advance investigations already initiated by others and to enable further inquiry in turn. The crucial question is how such dispersion is sufficiently overcome to enable such mutual dependence and to highlight some projects, practices, and achievements as especially significant in the context of that ongoing collective project' (1996, 167).

It is useful to contrast Rouse's answer to those he rejects. The coherence of science, he argues, is not explained by the 'standard view ... that scientific "communities" share fundamental presuppositions (vocabulary, theoretical models, values and norms, and so forth) which are inculcated in their members' professional training and which are enforced by the standard gatekeeping procedures of a discipline' (168).

Taking Rouse's contenders for explanations of the coherence of disparate scientific activities in turn, it might first be thought that the reason for coherence is that all genuine scientific knowledge can 'be captured in a single vocabulary, commonly [taken to be] that of theoretical physics' (176). According to this view, a scientific result gains significance by virtue of being described in terms that can in principle be reduced to a single form of description. Historically, a strong contender is the claim that what makes a phenomenon scientifically significant is that the proposition describing it can be reduced to propositions verifiable by the methods of the 'hard' sciences, such as physics. A contemporary challenger comes from biology: due to recent developments in genetics and biochemistry, the language of biology is championed as the successor to the language of physics as the primary form of description of scientific results (E.O. Wilson, 1998). A common form of description articulates relationships between new results and previous work.

Second, it might be thought that the coherence of scientific work is explained by a shared theoretical model, or paradigm. On this view, scientific results gain significance and connections to one another through their logical relations to theory, as, for example, confirming observations. Reliance on theory as the source of science's coherence has the effect of construing laboratory work as theory testing. Insofar as scientific theory is interpreted as a representation of the natural world, this view sees not Language, but Nature, as the basis of coherence. Third, coherence might be sought in shared norms and values in the conduct of scientific work. Merton's conception of scientific norms is the origin of this approach. Work violating the norms of science will fail to produce results that can be significant for previous results because they fall outside the realm of scientific work altogether. On this view, Society, not Language or Nature, is the basis of coherence.

The difference between Rouse's position and all three versions of the standard view is that each of the three see the significance of scientific results as grounded in a meta-context of shared presuppositions (Language, Theory, or Society) rather than in the labour of converting previous results into resources and configuring one's own work as a resource for others. The standard view posits a grounding context beyond the multifarious links that are actually forged through scientific labour. It is as if, even in the absence of such labour, the significance of scientific results and their relationships to one another would emerge simply as a consequence of shared presuppositions and internal features of the work itself. But for Rouse, coherence arises from a complex field of relationships embedded in the understanding of scientific endeavours as projections *from* a shared (but limited) context of past activities *to* a possible future in which current work can be taken up in likewise shared (but limited) research projects. He characterizes such understanding as the 'practical grasp of a research situation as a field of possible activities' (161).

Rouse's position has affinities with some already discussed. Scientists do not possess sources of unity and coherence of science by virtue of any particular scientist's or group of scientists' theoretical representations. Instead, each scientist or group of scientists possesses an understanding of local resources, projects, and imperatives governing their particular situations. By projecting previous work towards future collaborative work, they help construct a collective scientific field in much the same way as the various micro-processes described by Foucault interact to create coherent discursive formations. The fields thus created are not unified by anything other than the complex intertwinings of what might be called their sub-routines, not from any grand design. In Bazerman's terms, the scientist's 'schema' is not a shared conception of a single collective narrative but rather a specific narrative strand connecting one's own research to previous work and projecting it into the future as a resource for others. Bazerman's study of reading physics and Knorr-Cetina's concept of the journal article as a script of resource conversion imply that scientific literature is an important documentary resource for forging such connections.[21]

The claim that a practical understanding of the location of a research project in relation to shared contexts of past and future work has a narrative structure asserts that in acting as a scientist one already has 'some understanding of what it would be to have done the action in question. It is also to have some sense of how to initiate or continue the action now. This is in turn to have a grasp of the situation one is already in, to which the action is an intelligible response' (162). This kind of understanding situates action in a temporal sequence. It becomes embedded in a story or narrative that interweaves the significance of current research in possible futures projected from shared past histories. Disparate scientific activities cohere because 'the intelligibility, significance, and justification of scientific knowledges stem from their already belonging to continually reconstructed narrative contexts supplied by the ongoing practices of scientific research' (161). Contrasting his answer to that of the standard view, he says, 'I suggest that we take the coherence and significance of scientific work to be established by the ongoing reconstruction of contested narrative fields' (169).

Narrative understanding of scientific practices has three important features. First, in exhibiting a narrative structure, scientific research is not a unique kind of endeavour, but 'is rather like any other sort of human activity' (172). An explanation of scientific coherence therefore rests upon a general account of human action. Since scientific narratives are distinguished by specific contexts and problems, they differ from others only in the types of activities brought into relationships with one another. Scientific work is one cultural phenomenon among others. Activities not narratized in relation to activities already understood as scientific thereby place themselves beyond science, or at least as marginal to it. If, for example, scientists link their work to aesthetic rather than other scientific activities, drawing upon a context of previous and concurrent artistic activities and projecting towards a future of shared aesthetic creations rather than scientific concerns, then although their work may gain significance as a work of art, it has none as science, even if the very same results would represent a legitimate advance in knowledge if re-narratized.[22] Scientific achievements are informative and significant by virtue of their location in a narrative strand – a story – consisting of a temporally extended sequence of activities that already count as scientific.[23]

Second, in conducting their research, scientists find themselves (but only after years of training, passing many tests of their capacities and competence, submitting to a thorough discipline of social and professional relations, etc.) already in the midst of a particular field of possible research directions. In linking their work to a shared context of past and current work and projecting it onto future work through their practical deployments of scientific resources, scientists construct a story about a shared research context in which their work plays a leading role. It should not be thought to consist of the writings of scientific commentators, which is produced after the fact. Instead, it is located at the research front, in the way in which scientists construct meaning for their research. As Rouse puts it, the narrative they construct 'is not a story told in retrospect but a story that the narrator is in the midst of' (163).

Third, narrative understanding does not take the form of adapting one's work to a single, grand narrative of science.[24] Narrative labour is not the effort of positioning particular results in a single story. Instead, every scientist is in the midst of a specific story which has complex relations to others, including contests and conflicts. 'In the narrative fields of scientific practice,' Rouse argues, 'from no unitary authorial point of view can an entire course of events be surveyed, for multiple authors engage in an ongoing struggle to determine the configuration of the narrative within which they are situated.' The coherence of the scientific field consists in a culture of narratizing, rather than in the construction of a single product of such activity: 'Much of what happens in day-to-day science concerns the emergence of a coherent narrative field of action from the multifarious doings of different scientists, whose work aims to push the story line in different directions' (161).

Scientific narrative is the intermeshing of specific stories drawn from a shared field of narrative resources. The main problem with the standard view is that any account of the significance of scientific activity must account for conflict, disagreement, and changes in the structure and composition of scientific communities. Explanations relying on shared presuppositions allow no room for failures to share them. Yet studies of practices show that no presuppositions are in principle exempt from change, replacement, loss of interest, or outright rejection. A plausible answer to the question of coherence must make room for work which is scientifically significant even though it contests prevailing activities, phenomena, training, techniques, methods, vocabularies, theories, patterns of behaviour, communal commitments, and so forth. The concept of a narrative field can account for conflicts and meaningful differences. Narrative understanding constructs the identity of a scientific field 'only as a field of intelligible differences, one whose boundaries are themselves contestable' (173), not, as the standard view would have it, in terms of shared presuppositions. The coherence of scientific practices depends only upon a common terrain of action performed upon a shared narrative field that provides resources for contested stories rather than a single coherent story. 'What is held in common,' Rouse observes, 'among practitioners of scientific research is a field of interpretive conflict rather than any uncontested commitments about beliefs, values, standards, or meanings' (170–1). His inquiry into the nature of this shared field of action is perhaps the most thorough analysis of the resource connections between different centres of scientific practices.

Narrative and Documentation

If, as Rouse claims, scientific understanding is fundamentally narrative in structure, and given the materialist tendencies evident in his analysis of the importance to scientific practices of occasioned utterances and inscriptions, the question arises, how are specific document forms related to the narrative structure of scientific action? The journal literature is one such form. Another is the 'contingent repertoire' analysed by Gilbert and Mulkay. Laboratory 'shop talk' has been studied by Lynch (1985), who with Woolgar (1990) has also drawn attention to the role of pictorial representations. Lenoir (1999) has examined the transformation of biomedicine by information technologies. Several titles in the Stanford University Press series, *Writing Science*, explore ramifications of the 'materiality of communication,' especially Gumbrecht and Pfeiffer (1994) and Lenoir (1998). The connection between narrative and documentary types suggests further research problems: (1) Is it possible to develop a typology of institutional sites of scientific practices organized by unique or characteristic documentary forms? (2) How are the discursive resources made available by different documentary forms used in narratives that establish the meaning and significance of scientific results? (3) Do the differences between documentary forms and the resources they make available generate corresponding differences among narrative understandings of scientific results? Questions such as these form a subset of research problems concerned with the role of documents of various kinds in many different scientific fields.[25]

Rouse abstains from addressing questions about the role of particular forms of scientific documentation. For him, narrative understanding is characteristic of *action in general*, not of actions performed at specific sites, such as the scenes of scientific documentary practices. In his explanation of epistemic alignments, Rouse points out that local phenomena and results can be informative elsewhere through the work of standardization and routinization of tools, techniques, theories, and disciplined behaviours (1987, 69–126; 1996, 179–204). Its narrative form does not distinguish scientific activity from other practical and intentional activities, but is merely, so to speak, a particular cultural 'flavour' of action in general. Since narrative is located at a deep structure of intentional activity, the 'multiple, ongoing, narrative unifications of scientific knowledge [that] are part of everyday scientific research practice' (1996, 176) should not be reduced to the production of explicit statements about the connections between one's own work and work of others. Narrative understanding no more requires any such explicit statement or belief than my grasp of an opportunity to score a goal in a hockey game requires that I utter a statement or entertain a belief that my pass to the winger is a good move. My narrative understanding is enacted in the pass itself – in action, not in cerebration, or in talk. Similarly, scientists' understanding is enacted in the ensemble of moves they make, both discursive and nondiscursive, which situate them in an ongoing story of collaborative scientific research.

Rouse is therefore not concerned to locate narrative understanding in any particular form of articulation of scientific practices, even while acknowledging many such forms: 'Scientific discourse takes place at many levels, from informal conversation to formal presentations or publications and eventually to retrospective review and incorporation into reference works' (1987, 123). In locating the resources involved in the production of 'scientific knowledge as narratively reconstructed epistemic alignments' (1996, 176), Rouse appeals not to any of these specific documentary strata, but instead to what 'everyone knows.' Since narrative understanding takes the form 'not of an explicit representation but of a practical capability' (162–3), what 'everyone knows' is enacted in practice in general, whether material or discursive. Observing that what everyone knows 'often runs ahead of publication,' Rouse notes that it 'cannot be identified with any stage of the published literature' (1987, 123). He therefore says, 'I am not concerned with the form in which the results of investigation are written. I do not want to claim, for example, that scientific papers are really narratives in disguise' (1996, 160).

Despite Rouse's deliberate rejection of an equivalence between narratizing and production of scientific writing, his concept of narrative nonetheless does not rule out investigations of documentary types that might be shown to play specific roles in the enactment of scientific narratives. Since all scientific work manifests narrative understanding, no particular kind of labour, including documentary labour performed in producing scientific literature, is privileged as a narrative resource. Moreover, the cultural phenomenon of modern Northern/Western science is marked by an imperative to produce formal literature. Scientists need not only to 'make things work,' but also to produce inscriptions according to the strictures of institutional writing. The importance of particular documentary forms depends solely upon how their statements are related to other elements of scientific work. Knorr-Cetina has shown how the scientific paper derives from other writings, both the journal literature – the 'scriptures' of the field – and the written traces of laboratory work. Latour and Woolgar have shown that the journal literature is so thoroughly integrated into scientific research at the level of laboratory bench work by virtue of its centrality to scientists' shop talk that they regard the laboratory itself as a machine of literary inscription. Bazerman found that the schema of significance for scientific results of the physicists he studied also derive from the journal literature. Writing is not seen as an after-the-fact labour process, one easily detachable from the real work with instruments, tools, and the many material elements found at the laboratory bench. These studies imply that the journal literature is thoroughly intertwined with both other documentary forms and the many non-documentary practices of the natural sciences. Its discursive form bestows no privilege as a resource for narrative understanding of scientific practices. But its materiality grants it a claim equal to those of non-discursive resources.

Certain features of the narrative field of scientific practices invoke Foucault's concept of the materiality of statements as institutionalized patterns of occasioned inscriptions. One such feature is its enabling us 'to see how scientific work achieves coherence and significance in a way that allows for conflict and meaningful differences' (169). The tension between coherence and conflict requires that discipline be exercised. As Rouse puts it, the 'need to make differences intelligible and a common project possible compels an ongoing struggle to keep in check the divergence of versions of a community's story, even as the various actions of its members strain at the limits of coherent inclusion with one another. This struggle takes the form of a shared concern to construct, enforce, and conform to a common narrative pattern within which everyone's endeavors make sense together' (165). Moreover, the discipline is imposed by the subjects themselves, analogous to panoptic self-discipline: 'researchers preempt the need for discipline imposed from without by adjusting their work to (what they take to be) the prevailing epistemic situation' (173). Following Foucault, we can seek particular institutional exercises of discipline over particular discursive forms. The institutionalized production of science's formal literature is but one example. Knorr-Cetina has shown that the scientific paper is a highly disciplined discursive form which represents a resource conversion typical of scientific work. Thus science's institutional imperatives of inscribing scientific results in a highly disciplined journal literature yield a set of resources for the documentary articulation of scientific narratives. To materialize Rouse's concept of narrative in this way is not to claim that 'scientific papers are really narratives in disguise.' The rules of writing that constitute the institutional materiality of this literary form contribute a documentary technology to the narrative field of resources for constructing contested stories. Scientific writing is just one among many resources for the exercise of narrative understanding of scientific phenomena.

Studies of scientific practices therefore find an important role for the journal article at the laboratory bench. It is not the role of transmitting 'information' among rational, information-processing agents. The ways in which imperatives of writing configure laboratory labour should therefore be expected to take different, more complex forms. As narrative enactments, the inscriptions of formal scientific documentation constitute neither the theoretical representations nor the 'information' imagined by IC models. Rouse's comments about narrative understanding per se remind us of the non-representational nature of science's contested stories: 'These stories are not *about* our world but *of* it. They are not representations of events but the events themselves as significantly configured by their place within a contested story field' (163–4). The idealizations of epistemological conceptions of information seeking gain no purchase upon the complexities of science's narrative enactments.

The work of Shapin and Schaffer, Latour and Woolgar, and Knorr-Cetina show that the rules of writing governing the journal article provide discursive resources for the work of objectification. Such resources, together with the exercise of institutional discipline in their use, strip laboratory results and phenomena of their particularity and insert them in a shared resource field. The rhetorical product of such resource uses are objective facts viewed by 'modest witnesses.' The objectifying resources of the journal literature constitute a platform of articulation from which, as Latour has insisted, Nature emerges, simultaneously present to all, yet the privileged domain of a finely honed and institutionally protected scientific consciousness. Although historical research shows that things can be made to work and knowledge of the natural world can be gained without the kind of narratizing enacted in modern journal literature, studies of scientific practices show that the imperative to document scientific results in such a manner is an important feature of the historically and culturally contingent phenomenon of modern Northern/Western science.

Although a very useful resource for thinking about the configuration of scientific research in contemporary scientific practices, Rouse's concept of the narrative structure of scientific understanding must be treated with caution when analysing scientific work in other times and places. Because he grounds narrative understanding in a general account of human action – specifically, the structure of *Dasein* and purported 'needs' to align one's work with others – his analysis operates on a level of generality ill-equipped to articulate historical differences, and in particular to illuminate instances of knowledge production that depend less on shared narrative fields than does contemporary scientific work. Even if one agrees that in each historical instance *Dasein* projects itself towards a 'for-the-sake-of-which,' we still need to know the form such practical narrative understanding takes in each instance. Examples of narrative understanding reveal only specific resources that enact narratives in historically contingent narrative fields. The resources of institutionalized documentary regimes vary widely in different times and places. To investigate the differences between them, an analysis more finely grained than the universal structure of *Dasein* is needed. Is there a better reason to believe that science *in general* is grounded any more firmly in the structure of *Dasein* than in Nature, Language, or Society?

In the course of his criticism of Heidegger's view of science, Rouse appeals to the materiality of scientific writing as an explanation of the decontextualization that privileges theory production. Interpreted as a statement about a particular historical manifestation of the organization of scientific labour, his claim is insightful because it insists on the material form of modern scientific writing as central to such decontextualization. He says: 'Scientific research, with its various and complicated activities in the laboratories, aims toward an outcome that can be shorn of all reference to the particularities of its production and situated in a context of representations (i.e., other inscriptions) that can be made entirely thematic because its background (the page) has been emptied of all content' (1987, 78–9).[26] The 'page,' however, has many historical manifestations – manuscript, print, electronic – and the printed form at issue in Rouse's remarks appeared much later than many robust forms of scientific knowledge. The objectifying resources manifested in a literary technology of science that printed inscriptions on a blank, content-free page are historically and culturally contingent. It is not essential to the production of scientific knowledge that facts be constructed in an abstract information space, a process in which specific technologies of document reproduction are implicated. In Traweek's terms, science's narrative understanding *need* not construct a 'culture of no culture' (1988, 162). Practical narrative enactments can, for example, consist in master-apprentice craft relations in which scientific results are made available for others but without the narrative form of the construction of objectified facts featured in the formal literature of modern science. Even modern engineering embodies knowledge used to 'make things work,' but it is to a great extent independent of the decontextualized information space constructed by the literary technology of scholarly scientific literature.[27] Reflecting on the durability of Roman bridges compared to those produced in eras of formal scientific writing, we may be motivated to find many other examples of scientific knowledge produced in different historical and cultural circumstances, of things having been made to work, yet without the objectifying narratives of our journal literature. Chapter 6 reveals some of the historical mutations of objectivity even since Robert Boyle's day.

Stratification

To accept scientific writing as an objectifying resource raises questions about the distribution of objectivity over science's several documentary sites. The literary technology of science is stratified, consisting of many layers, from the inscriptions and notes of the laboratory to conference papers, through the journal article's many drafts to its final, published version, and the subsequent incorporation of results in reviews, handbooks, and the many other forms of secondary and tertiary literature. Robustness of facts, bias towards certainty, and distance from ideals of objectivity will vary with the documentary platform of enunciation. The degree to which the objectifying project is realized depends not on the number of truths articulated at particular levels of scientific writing. As Rouse points out, not only is a complete representation of the natural world absent from scientific documentation, but even the most apodictic statements of the tertiary literature are open to revision by virtue of the non-representational logic governing their selection:

> neither the scientific literature as a whole nor any of its component parts can be understood as an encyclopedic representation of the present state of knowledge in the sense in which knowledge is typically reified by most epistemologies. The journal literature is obviously not an attempt to represent the present state of scientific knowledge, for it includes contested and conflicting claims. But the various places in which knowledge might plausibly be said to have been collected and reviewed (for example, review articles, handbooks, textbooks, popularizations, or encyclopedias) will not do either. Such collections are always put together for the sake of

some subsequent use, which governs the selection and interpretation of
its contents ... What the books are about and hence their contents and
organization are at issue in ongoing scientific practice. (1996, 174)

Arguments for the temporal and deferred character of knowledge
rule out construing objectivity – at any particular level of scientific
writing – in terms of contributions to theoretical representations of the
natural world. Thus the 'information' that epistemological models
assume is conveyed at particular documentary strata is not the source
of objectivity. Instead, the objectifying resources of documentary systems consist in the kinds of statements made available, the standards
of selection applied to them, and the degree of occlusion of particularity, situatedness, locality, and context. As the modality of statements
moves from the hypothetical to the apodictic, and as the projected referents of the terms used in statements tend towards the universal, the
degree of objectivity manifested through writing increases.

Objectivity by documentary means is an ongoing *project*, not the
achievement of a completed representation. It is always under construction, but never fully achieved. The objectivity of the natural world
is an effect of its deferral – a correlate of the open-endedness of the
narrative structure of scientific understanding, the persistence of modern science's objectifying imperative and the continual institutional
discipline exercised over its objectifying resources. Science's literary
technology offers a stratified set of discursive resources deployed in
processes of standardization, routinization, and occlusion of locality,
context, and contingency. These processes mobilize both documentary
and non-documentary elements. Insofar as the ojectifying project is
carried out through writing, it consists in a rhetorical strategy of constructing a discursive space from which statements emerge only as *candidates* for the stability facts enjoy. The strategy of scientific writing
gives Nature a voice, not a script.

The complex link between objectivity and scientific writing is
addressed in Fleck's (1979) discussion of the levels of scientific documentation corresponding to the stratification of scientific knowledge.
His reflections on 'the genesis and development of a scientific fact' are
among the most useful of any in understanding the role of writing in
objectification.[28] He begins by dividing a scientific community, which
he calls a thought collective, into esoteric and exoteric circles. The esoteric circle consists of the group of scientific experts active in a particular research area; the exoteric circle consists of 'educated amateurs.'

Based upon this distinction, scientific knowledge is divided into expert and popular knowledge. The esoteric circle is subdivided into specialized and general experts. Specialized experts are 'informed at the greatest depth – for example, a radium specialist in the science of radioactivity' – and are located at the centre of their circle. The general expert includes 'scientists working on related problems – all physicists, for example' (110). The distinction between specialized and general experts is reflected in the distinction between '*journal science* and *vademecum science*, which together constitute expert science.' A fourth category which Fleck does not analyse further is *textbook science*, 'based upon special methods of teaching' (111).

Fleck analyses popular, journal, and vademecum science in terms of the degree of certainty of statements. Certainty increases as a stratum of scientific documentation moves away from the site where knowledge is manufactured: 'The greater the distance in time or space from the esoteric circle, the longer a thought has been conveyed *within the same thought collective*, the more certain it appears' (106). Popular science is the most apodictic of the three; it aims at certainty, simplicity, and vividness. Its appeal and force depend on the 'omission both of detail and especially of controversial opinions' (112). The scientific terms it employs seem to refer to stable, objective entities because its literature shows no trace of the labour involved in the construction of epistemic alignments. Its descriptions, Fleck says, read 'as if definitive concepts and ideas existed a priori' (116). As the literature moves to the 'grossly popular' from the merely popular, 'thinking appears to be even more strongly dominated by an emotive vividness that imparts to knowledge the subjective certainty of something holy or self-evident' (117). 'Simplified, lucid, and apodictic science,' says Fleck, 'are the most important characteristics of exoteric knowledge' (112–3).

Popular science is not merely derived from esoteric science. Fleck claims that it furnishes 'the major portion of every person's knowledge.' Scientific experts are not excluded, because 'even the most specialized expert owes to it many concepts, many comparisons and even his general viewpoint' (112). The 'general viewpoint' of the expert consists in a particular *Weltanschauung* whose distinguishing feature is the standpoint of objectivity. Popular science is epistemologically significant for precisely this reason. It provides 'the background that determines the general traits of the thought style of an expert. This may sometimes be no more than an exalted feeling about the solidarity of all human knowledge. Or it may be a belief either in the possibility of a universal science or in the albeit limited potential for further development in science' (113). Popular science therefore yields two distinct forms of objectification: a general orientation towards objectivity, and specific concepts and ideas taken as 'given a priori' (in Latour's terms, these concepts and ideas have been 'black-boxed'). The expert's faith in the certainty, simplicity, and vividness of science derives from popular science. The form of scientific certainty is materialized most concretely in its documentation. Its simplified, vivid, and apodictic statements provide the framework for scientific practices, including those of the research front. Fleck's claim that popular science reacts upon the expert can be read as an argument for the workings of the most vivid of objectifying resources – and perhaps the most unexpected – at the level of highly specialized scientific work. Popular-scientific literature functions both as a constraint and an opportunity: as a constraint, because results not stripped of locality and contingency are unacceptable to the thought collective, and as an opportunity, by specifying a particular configuration for acceptable results.

Vademecum science is a site of transformation, where scientific statements gain a higher degree of certainty than in the journal literature, where they are fragmentary and tentative. Science in its vademecum (or handbook) form is characterized by 'a *critical synopsis in an organized system*' (118). Fleck says that a 'statement appears ipso facto more certain and more soundly established in an organized system of a discipline as presented in a vademecum than it does in any fragmentary description found in a journal' (121). Like narrative reconstructions of scientific activities, the plan of the vademecum emerges from a contested field: 'Such a plan originates through esoteric communication of thought – during discussion among the experts, through mutual agreement and mutual misunderstanding, through mutual concessions and mutual incitement to obstinacy' (120). But unlike Rouse's conception of narrative, Fleck's vademecum aims at a closed system: 'A vademecum is built up from individual contributions through selection and orderly arrangement like a mosaic from many coloured stones ... When two ideas conflict with each other, all the forces of demagogy are activated. And it is almost always a third idea that emerges triumphant: one woven from exoteric, alien-collective, and controversial strands' (119, 120). Collective work yields facts that already belong to a system, one which manifests the thought style of the thought collective. The vademecum is therefore governed by its 'plan.' Once constructed, it plays a role similar to scientific narrative, but by virtue of a fixed story rather than by a collective but contested narrative field: 'The plan according to which selection and arrangement are made will then provide the guidelines for future research. It governs the decision on what counts as a basic concept, what methods should be accepted, which research directions appear most promising, which scientists should be selected for promising positions and which should simply be consigned to oblivion' (119–20). The significance of any element of the vademecum – such as the etiological concept of disease entity – derives from 'a directed selection of individual investigations and a directed compilation ... once part of the vademecum, it is taught and generally used. It forms a keystone of the system and thus exerts a constraint on thinking' (121).

Vademecum science is the level of scientific activity where facts are produced. But they are produced by collective labour, not by individual effort. The particular fact studied in Fleck's book provides an example. The precise origin of the connection between the Wasserman reaction and syphilis is elusive because it derives from convergences of the many micropractices of many different workers. 'No authors can be specified who consciously brought it about. We cannot state exactly where it occurred nor explain logically how it happened' (72).[29] No fact can emerge without the kind of *disciplined practice* that made the reaction useful, a result brought about with 'disregard for theoretical questions and the ideas of individuals' (73). Social discipline is required for the stabilization of scientific phenomena. Fleck's study makes this point very emphatically:

> Wasserman's reports about his reaction contain only the description of the relation between syphilis and a property of the blood. But this is not the most important element. What is crucial is the *experience* acquired by him, by his pupils and in turn by theirs, in the practical application and effectiveness of serology. Without this *experience* both the Wasserman reaction and many other serological methods *would not have become reproducible and practical* ... even today, anybody performing the Wasserman reaction on his own must first have acquired comprehensive experience before he can obtain reliable results. Only through this experience will he participate in the thought style, and it is experience alone that enables him to perceive the relation between syphilis and blood as a definite form. (96)

This passage can be read as the insufficiency of the communication of information to the fact construction. Wasserman's reports communicated the appropriate scientific information. But without the disciplined practical experience of performing the Wasserman reaction, such 'information' has no effects beyond those of patterns of ink on a page. Collective effort produces vademecum science, but without guidance from 'information,' the intentions or ideas of specific individuals, or from theories. The fragmentary contributions of many individuals and groups circulate throughout the thought collective and in the process are transformed, thereby enabling the kind of disciplined experience and practical skill that make it possible for a fact to emerge and stabilize. When practical action is socially disciplined in the manner of the refinement of the Wasserman reaction, a context is created in which the statement documenting the fact can function as a thought constraint on further collective work. The Wasserman reaction becomes part of 'collective, generally valid vademecum science' (120); it becomes a taken-for-granted resource for further work.[30]

Beyond his use of the term 'vademecum' itself, which suggests the handbook literature, and his claim that it is not identical with the journal literature, Fleck is not concerned to specify further the documentary site of vademecum science. Vademecum documentation inscribes the accepted statements of the thought collective (e.g., 'The Wasserman reaction is a test for a property of the blood indicating the disease entity syphilis'). Through a stylization characterized by their degree of certainty, they are positioned in an organized system. Fleck tells us no more than this about the documentary site of vademecum science. The most likely candidates for its documentary form are the many manuals and handbooks used in a broad range of scientific work, from teaching to the laboratory bench. Yet because this literature fails to represent, as Rouse points out, the current state of scientific knowledge, it therefore fails to exhibit the 'plan' of the vademecum. The directed selections and compilations involved in the production of handbooks and manuals are undertaken for specific and limited purposes. As those purposes change, handbooks are altered and reissued. Ravetz notes another problem of this literary form, stemming from its lack of currency: 'the uncritical use of handbook information, which is always incomplete, obsolescent and not quite fitted to the needs of the work at hand, can lead to the most astonishing blunders' (90). Handbooks and manuals are good sources of evidence for the organization of thought in particular historical periods. But their selections and structure do not derive from and therefore reflect the impersonal mechanisms Fleck takes to be at work in the production and organization of vademecum science. Furthermore, studies of practices show that the organization and intelligibility of scientific results do not derive from the kind of stability and unity Fleck assumes in his analyses of thought styles. Scientific activity does not produce Fleck's stable mosaic; it is more like a process aiming at its construction, but which is always contested and whose pattern is always open to revision and variation. Fleck's failure to specify the documentary form of the vademecum may reflect an idealization of vademecum science itself.

Journal science differs from vademecum science because its statements are provisional, individual, and personal. Fleck claims that it lacks the critical synopsis of an organized system. Journal writing does not represent 'the migration of ideas throughout the collective' (119). It consists instead of 'various points of view and working methods,' 'so personal that no organic whole can be formed from the contradictory and incongruent fragments' (118). Not only are its problems fragmentary, it is also characterized by 'the contingency of the material ... the technical details, in short, the uniqueness and novelty of the working material' (119). Although the scientific paper displays the caution and modesty noted by later commentators (Fleck calls the substitution of 'we' for 'I' in scientific writing 'the plural of modesty,' 119), he interprets these literary devices as indexes of the fundamentally personal character of the paper, which they unsuccessfully attempt to camouflage. And, although 'any paper published in a scientific journal contains in the introduction or conclusion ... a connection with vademecum science,' this should be regarded 'as proof that the author aims at incorporating his paper in the vademecum and regards its present state as provisional.' 'It is not possible,' he concludes, 'to produce a vademecum simply from a collection of articles that have appeared in journals' (118).

From the perspective of studies of practices, Fleck's account of the journal literature is problematic. Knorr-Cetina has shown that the journal article represents a highly impersonal conversion of reason whereby the thoroughly contingent and localized opportunistic reasoning governing resource selection is completely decontextualized. Furthermore, the paper's results are recontextualized in a script of general and collective resources that are often beyond the scientists' own field of activity. At least some of the objectification Fleck locates in the vademecum is therefore already present in the journal article. Furthermore, Latour locates the organization of the journal literature in the journal literature itself rather than in a secondary literature derived from it. Such organization is evident not only in strategies of altering modalities of statements but also in the effects of the uses others make of journal articles. The critical synopsis of an organized system is therefore a goal of labour processes internal to the journal literature, even though the organized system is never complete. Finally, Rouse argues that what is shared is a narrative field than a common story. Insofar as the vademecum exists in some stratum of scientific documentation, it suffers the instabilities of production for a variety of subsequent uses. The objectivity of Fleck's vademecum might better be sought in shared narrative fields of objectifying resources than in critical synopses of organized systems. The great benefit, however, of Fleck's working through the implications of his claim that the 'thought style may also be accompanied by a technical and literary style characteristic of the given system of knowledge' (99) is to draw attention to the ways in which the stratification of objectivity, from lesser to greater apodicity, is manifested in a corresponding stratification of scientific documentation.

Conclusion

The writers whose work has been featured in this chapter point in new directions for thinking about the role of science's formal literature. Latour, Ravetz, and Latour and Woolgar emphasize the literary labour involved in scientific knowledge production – for Latour and Woolgar, scientists are 'manic writers' (1986, 48). The concept of modifying the modalities of statements is useful because it reveals how the work of scientific writing is performed both within the literature and in the laboratory. The latter is especially important, because it situates the imperative to produce stable statements in an institutionally authorized documentary form at the research bench. Knorr-Cetina's study of the production of a typical journal article shows how distant it is from a vehicle for conveying 'information.' Her concept of the article as a script of resource conversions actively sought in the laboratory also locates formal writing in the midst of the research action, but in a role very different than that imagined for information by epistemological views of science. Shapin and Schaffer investigate scientific objectivity as an artifact of discourse and literary technologies; Knorr-Cetina reveals its literary form: an abstract, Ravetzian information space. Bazerman's study of physicists analyses the thoroughly documentary nature of the 'schema' they employ to attribute meaning to research work. Rouse's concept of the narrative understanding of scientific action discloses a contested field of discursive and non-discursive resources used to project current work onto competing versions of a shared future. Fleck's analysis of the strata of scientific documentation describes differences in their contributions to the stabilization of scientific phenomena.

The analytical resources for understanding the documentary practices and circuits that have been extracted from this body of work emphasize their constitutive effects, rather than interpretations in terms of conduits for communication of epistemic content thought to be required by research science. A document becomes informing not by virtue of transmitting its message through documentary circuits, but due to the intertwined, institutionally disciplined, documentary and non-documentary practices from which 'information' emerges as an effect. The importance of the journal literature is not found in the truth value of its statements, but rather in the work they perform to increase the inertia – the weight and mass – that constitutes the stability of scientific phenomena. The resources of scientific documentation are not ethereal, epistemological, and conceptual, but material: they belong to ontological categories of substance, or things, not those of epistemology, such as thoughts, concepts, theories, ideas, knowledge or propositions. Views of information as epistemic content, of individual meaning or sense making, of knowledge gaps and repairs to incomplete world pictures or mental models of individual subjects, draw upon these mentalistic and epistemological discursive resources.[31] But if the labour of experimental science involves constructing material stability in the laboratory from a wide range of disparate elements, submission to social disciplines and institutional imperatives, and constructing large-scale network stability beyond the laboratory, the journal article does not become irrelevant to scientific work by failing to communicate, convey, or otherwise bring to consciousness the information thought to be required for research science. Instead, it contributes to the stabilization of networks of heterogeneous elements with and upon which scientific labour is expended. The journal article is best understood by abandoning the macroconcepts of information and communication in favour of a focus on the micropractices – scientific, social, and cultural – in which it is embedded. The next chapter explores the implications for documentation of various practices through which documents become informing.

6

Documenting Universality

This book's focus on the formal literature of science is intended to challenge epistemological presuppositions of information studies generally, not simply assumptions guiding studies of scholarly communication among scientists. Because the truths and facts thought to be conveyed by scientific documents are interpreted as paradigm cases of robust and stable 'information,' then, should the model of 'information seekers' pursuing and communicating epistemic content fail in the case of scientists, corresponding models become even less plausible for non-scientific professionals, much less for ordinary people pursuing their quotidian activities in which documents of various forms play such meaningful roles.[1]

Previous chapters have shown that the contingent, local character of laboratory resources and results is one of the most important findings of recent studies of scientific practices. Yet, as we have seen, scientific results are articulated as phenomena that are informative in contexts beyond those of their production; they speak with the voice of nature, not humanity. Traweek's studies of high-energy physics communities reveal 'an extreme culture of objectivity; a culture of no culture, which longs passionately for a world without loose ends, without temperament, gender, nationalism, or other sources of disorder – for a world outside human space and time' (1988, 162). In closing his study of the genesis and development of a scientific fact, Fleck writes about 'a common *reverence* for an ideal – the ideal of objective truth, clarity, and accuracy. It consists in the *belief* that what is being revered can be achieved only in the distant, perhaps infinitely distant future; in the *glorification* of dedicating oneself to its service; in a definite *hero worship* and a distinct *tradition*' (1979, 142; emphasis in original). Elsewhere, he refers to 'the harmony of illusions' which 'generates a firm belief in a reality existing independently of us' (87). Rheinberger notes that 'once a surprising result has emerged and has been sufficiently stabilized, it is difficult to avoid the illusion of a logic of thought and even a teleology of the experimental process' (1998, 290). Rouse's notions of epistemic alignments and the standardization and routinization of resource extensions also help explain the naturalization of localized scientific phenomena. Important too are Hacking's (1992) and Latour's (1988) observations about the reproduction of laboratory set-ups in other research contexts and transscientific fields.

When studies of scientific practices show that contingency is central to them, nature's objectivity and the universality of scientific claims cannot be taken for granted but instead need to be explained. 'One of the central implications of this work,' remarks Lenoir, 'is that universality must be constructed. An account must be given of how meanings locally produced are multiplied in other sites and how representations circulate and acquire global legitimacy' (1997, 18). The construction of universality as the cultural meaning of scientific phenomena and the naturalization of scientific representations were not problems for representationalist and theory-dominated models of science. For them, Lenoir points out, the 'truth of theories was a function of their fit with an independent world of objective facts. Appeals to the laws of nature and the assumption that rationally constructed representations mirror the structure of the world guaranteed the universality of scientific claims' (18). Commitments to representational knowledge of the natural world are not the only resources used to construct the universality of scientific claims. The social constructivist's appeal to the stability of social identities provides an alternative route to the same ideals of universality and objectivity by underwriting the stability of scientific knowledge with assumptions about the coherence of scientific communities. For epistemological realists, the universality of scientific knowledge simply articulates the stability of the natural world. For social constructivists, the objectivity of scientific knowledge rests upon stable communities, thus reflecting the stability of the explanatory categories of social constructivism itself (Rouse 1996, 245).

For Sandra Harding, the problem presented by the phenomenon of the universality of modern science is how to explain the success of an illusion. She argues that the universalizing project of modern science originated in European expansion and colonialism, which had several benefits: the testing of hypotheses 'over vastly larger and more diverse natural terrains than could other cultures,' the incorporation of elements of other cultures' 'ethnosciences' into European science, and the European destruction of many competitive local knowledge systems. She sees these strategies as 'the dissemination of a predatory conceptual framework for and by European sciences' (1998, 179–81), a framework she describes as 'the persistent substitution of abstract, transcultural and ahistorical concepts of nature and processes of gaining knowledge for concrete, locally situated, and historical ones' (181). Her work shows the stark differences in the cultural meanings of scientific results: the cultural form of European science is and has been a universalist and predatory 'culture of no culture.' Moreover, she acknowledges the role of documents and documentary institutions in the universalizing project: 'the cognitive content of sciences is shaped by culturally different forms of the social organization of scientific research. Scientific research is social labour, carried out in culturally distinctive kinds of organisations – laboratories located in industries, universities, physicians' offices, federal institutes or computer-connected collections of such sites; field stations, farms, collecting and observing expeditions, *conferences, learned societies, journals* ... and so on' (69–70; emphasis added).

To appreciate the cultural significance of documentary practices it helps to consider more closely the universalizing project's historical contingency. To that end, this chapter reflects upon various manifestations of the cultural phenomenon of scientific universality. With its contingency more firmly established, the cultural status of its supporting concepts – such as epistemological concepts of information and communication – is more clearly evident. The conceptual framework of what follows rests upon my claims that (1) a shift to understanding science as a historically contingent form of cultural production implies a corresponding shift in studies of scientific documentation; and (2), when institutionalized and socially disciplined practices involving the various material forms of occasioned utterances and inscriptions, together with their production, appropriation, and various cultural articulations replace information and communication as privileged analytical tools, studies of effects of documentary practices and circuits provide fruitful alternatives to the epistemological interpretations of documents currently favoured in information studies. The final section of the chapter discusses five analytical resources which, I argue, are important tools for documentation studies.

Contingent Universality

Perhaps the most direct route to an appreciation of the historically contingent, hence tentative and contested, character of the universality claims of modern science is through cases of things and processes having been made to work, and of success in prediction and control over natural phenomena, but which do not exhibit the universality characteristic of modern science. With such cases in mind, we can more easily see that the production of the universality and objectivity of scientific knowledge is a historically contingent phenomenon. The sort of cases we need come from premodern science, science that is for different reasons marginal to European science, and contemporary cases of 'making things work' that fall outside the culture of scientific research.

The History of Universality

Harding's work reveals that many 'exotic' ethnosciences were absorbed into the European scientific project rather than left to develop according to their own non-universalizing traditions. The local knowledge systems of premodern Egypt, Greece, pre-Columbian civilization, India, Arabic cultures, China, and marginalized European traditions reveal the cultural diversity of scientific production when judged by criteria of successful prediction and control of natural phenomena.[2] Furthermore, if the products of 'exotic' and marginalized local knowledge systems can be appropriated by European sciences, universality is not necessary to knowledge production generally. That the same phenomena can be produced in both cultures, that the same predictive power and control over nature can be exercised in each, and yet the phenomena produced in the colonized culture count as scientific only when absorbed by European scientific institutions, indicates that the project of universality is a highly contingent, politically charged, and historically and culturally specific enterprise.

Not only is universality not necessary to scientific knowledge, but the history of the European scientific tradition reveals that universality itself is subject to historical mutations. In his investigations of 'the mathematical way in the scientific revolution,' Dear (1995) shows just how much documentary labour has to be performed to control threats to institutionalized forms of universality. His concern is how the continental philosophers of the early seventeenth century responded to the conflict between the Aristotelian deductive route to scientific truth and the advent of modern experimental methods, techniques, and instruments. For Aristotelian physics, the universality of knowledge claims took the form of reference to the essential natures of things. The typical knowledge claim was a 'statement of *how things happen* in nature, rather than a statement of *how something had happened* on a particular occasion' (Dear 1995, 4; emphasis in original). Before the advent of experimental science, '[e]xperience taught how nature usually behaved; it did not consist of knowledge of discrete events, because such events might be anomalous, "monstrous"' (21). Thus the universality of knowledge claims was grounded in 'how things work' without the later assumption of laws governing the behaviour of all natural phenomena. 'Monsters' were possible in nature, because 'the physical world was a concatenation of established but sometimes wayward rules, not a logically integrated puzzle' (4).[3]

A typical statement of 'how things work' would be 'the sun rises in the east,' a statement based upon repeatedly confirmed observations available to any competent witness. Singular and unusual events, the kind that play such an important role in modern science's 'crucial experiment,' were not revealing of nature's behaviour. Instead, regular observations, 'commonly accepted, either through daily familiarity or through the statements of a weighty authority' (22), provide the certainty required to warrant universal knowledge claims. Such claims were statements about how phenomena *typically* behave. Given their certainty, they could enter into syllogistic reasoning aimed at demonstration or theoretical description and explanation. Since real knowledge was of essences, or universals, no immediate perception of a singular event could be epistemically significant. And since the observed regularities of nature could not have the certainty of mathematical demonstrations, duly legitimated and trusted authority was critical for the statements of natural philosophy. Thus for Aristotle, 'the nature of experience depended upon its embeddedness in the community; the world was construed through communal eyes' (23). For the kind of universality found in continental European premodern natural philosophy, '[e]xperience and social accreditation were never sharply distinguished' (23).

The seventeenth century confronted Aristotelian methodological practices with new ways of making knowledge. As natural philosophers developed the sciences of optics, astronomy, and mechanics, they often relied upon special instruments and techniques. The development and use of such tools, together with a weakening of the social structures of academic knowledge, meant that nature was increasingly observed by experts rather than sources of communal authority. The observations of interest to natural philosophers were not available through daily familiarity because they were technologically mediated through specialized apparatus. Nature had become less accessible to common assent. The 'experiences' upon which the developing science of optics, for example, came to depend 'were only truly evident to specialized investigators operating with tools designed for the purpose' (59). Astronomy too, relying as it did on the use of specialized instruments on specific occasions by expert observers, posed the problem of grafting universality – previously grounded in community – onto new kinds of experiences available only to a few: 'This was empirical knowledge that depended on the diligence and expertise of observers and experimenters, with their specialized instruments and skills; it was not a straightforward matter for others to reproduce for themselves the experiences claimed by astronomers or opticians. How could "experiences" be established as common property if most people lacked access to them?' (59).

The problem can be posed anachronistically as one of how to express these new kinds of experiences in statements that can possibly be *informing*.[4] Stated in this way, it is a problem not of communicating information but of constructing it. Given Aristotelian practices of fabricating universality, not just any statement – no matter how faithfully it might 'communicate information' about a singular event – could be accepted as a valid knowledge claim: '"Experience" in scholastic philosophy and mathematics ... typically took the form of universal statements because singular statements, statements of particular events, are not evident and indubitable, but rely on fallible, historical reports ... Singular experiences were not public, but known only to a privileged few; consequently, they were not suitable elements of scientific discussion' (44).

Dear's analysis of the natural philosopher's solution to the problem relies upon his insight that experiment and its documentation are closely linked. He sees scientific literary practice as 'a crucial feature of scientific practice as a whole ... an account of an experiment is an essential part of its performance' (1991, 135). Institutionalized documentary practices help manufacture credibility of documents. In early seventeenth-century continental Europe, 'epistemology, method, and text were woven together in the assumptions of that dominant scholastic pedagogy which took Aristotle as its touchstone of legitimacy' (137).

The documentary task was clear: purported 'observations' yielded by experiments and new instruments had, somehow, to be documented as Aristotelian 'common experiences.' To put it once again in anachronistic language of information, the trick was to present observations gained by means of contrived experiment or specialized expertise as scientifically *informing*, in a scientific culture where the criteria of informativeness are derived from common rather than singular experience. The key feature of common experience was that it was *evident*. If singular experience could be made to appear as evident as common experience, and for much the same reasons, the observations available to a few could be seen as continuous with those available to all.[5]

One of the literary techniques used to constitute a singular experience as scientifically significant was to strip its documentary account of any trace of contingency. The reported experience could then be presented with the *typicality* of evident, common experiences – of how things happen in nature. Thus when Roderigo Arriaga in 1632 reports his experiments on falling bodies, he 'does not find it necessary to provide a narrative of a discrete historical event' (Dear 1995, 69). The problem was especially acute, because common experience held – contrary to what Arriaga, like Galileo, discovered – that heavier bodies fall faster than lighter ones. Had Arriaga presented his finding in the form of a discrete event, there could be no way to demonstrate it was not a 'monster.' Instead, he 'states that he has tried [the experiment] "not once or twice, but often," so that his account is explicitly subsumed to a normal state of affairs known to him through experiential familiarity' (68). The form of his claims are universal – 'I find this,' rather than 'I found this' – and his description of a test serves only as an example of what typically happens. The order of statements in modern science, in which the specific case supports a general claim, is reversed: in premodern science, the general, universal claim is primary, with all specificity of discrete trials erased such that only the typicality required to show that the universal claim is *evident* remains. In presenting his account in such a manner, Arriaga appeals to Aristotelian cultural conventions of universality. His 'discussion is designed to create conviction, or at least a willingness to lend credence, in readers who have certain expectations about how experience ought properly to be construed and utilized in the establishment of a knowledge-claim' (69).

A second literary device consisted in borrowing the documentary form of the most evident knowledge claims available in the premodern period: the geometrical demonstration. The outcomes of such demonstrations were indisputable because they flowed directly from geometrical constructions used in proofs. Could the outcomes of experiments be made as evident by making them flow as directly from experimental procedures? The conventions of making mathematical statements evident functioned, in the documentation of premodern experimentation, as resources for rendering statements about natural phenomena as nearly evident as geometrical demonstrations. Thus in many accounts of experimental procedure, 'the series of operations is presented in the subjunctive mood of a geometrical *problema*' (59). The effect was to graft the prevailing cultural form of universality onto empirical results by blurring the distinction between them and geometrical proofs: 'A geometrical construction is transparent, because in following its steps one sees the outcome generated inevitably before one's eyes; even recourse to compass and ruler is unnecessary ... the geometrical literary structure did ... serve to accord a sort of transferred transparency to described experimental procedures. Procedure and outcome appeared formally inseparable ... Just as constructions in geometry were generated from postulates that expressed conceded possibilities, so the use of a geometrical paradigm served to re-create unfamiliar experience by generating it from familiar experience – that is, easily picturable operations' (60).

This geometrical documentary form made statements informing about nature not by virtue of their communicative effects – not by communicating information – but through their constitutive effects: the universality required of credible, evident statements of how things happen in nature had to be *made*. The authors of these accounts, by using the cultural resources available to them, and their readers, who used the same resources to interpret the accounts, were not 'information seekers,' nor were they engaging in 'information behaviours.' They were constructing statements – in the face of challenges from newly minted expertise and new instruments – from the historical and cultural resources that made it possible for their statements to be informing, in accord with the dominant, historical and cultural form of universal knowledge claims: Aristotelian 'axioms' of *how things happen in nature*.

Objectivity took a different form in Restoration England. Dedicated to a rigorous Baconian program of attention to singular facts, Boyle and his followers cast a healthy suspicion on universal generalizations and eschewed the sources of authority that supported them. In so doing, they faced a different kind of problem than their predecessors and continental counterparts, who relied on Aristotelian and mathematical resources to construct universality for experimental phenomena. 'The probabilistic model of Boyle,' Dear writes, 'required a category of the "matter of fact," the legitimacy of which depended precisely on accredited, and therefore specifiable, occurrences' (1991, 162). Continental and English practices were radically different: 'Boyle did not use axiomatic deductive argumentative structures, which were supposed to constitute science in an Aristotelian sense, whereas Galileo and the Jesuits did. The difference is of great significance. Boyle reported singular historical events; they needed universal statements of behavior even when giving historical accounts by way of collateral' (162). To the continental philosophers, a 'historical report of a specific event, of the kind that Boyle wrote endlessly, would have been scientifically meaningless; it would have been philosophical antiquarianism' (1995, 209). For Boyle and his followers the problem was how to generate credibility for statements of singular, historical, experimental events – without relying on extra-experimental assumptions used by their continental counterparts to support statements of events as typical of how things happen in nature – which nonetheless present robust, singular facts with the objectivity required of the observation statements of inductive science. True to their Baconian commitments to building knowledge from certified occurrences, at particular times and places, of granular, theory-resistant *facts of nature* rather than from axioms or 'essences,' they saw Aristotelian accounts of scientific results as a surrender to discredited authority. Rejecting certainty as the standard of natural knowledge, they adopted the more modest criterion of high probability. Shapin describes the strategy: 'Physical hypotheses were provisional and revisable; assent to them was not necessary, as it was to mathematical demonstration; and physical science was, to varying degrees, removed from the realm of the demonstrative. The probabilistic conception of physical knowledge was not regarded as a regrettable retreat from more ambitious goals; it was celebrated by its proponents as a wise rejection of failed dogmatism. The quest for necessary and universal assent to physical propositions was seen as improper and impolitic' (1984, 483).

Shapin and Schaffer's (1985) work shows that Boyle addressed the problem of conferring credibility on experimental experience by striving to create and maintain a community of experimenters – the Royal Society – whose gentlemanly civility and modest witnessing functioned as marks of trust in claims that experimental phenomena were indeed what they were reported to be. With the social signifiers of credibility in place, this community of experimenters can successfully attest to the natural-philosophical significance of singular facts in the name of a Baconian science in the making. There was no need to worry about the 'monsters' that Aristotelian natural science allows nature to breed because, once reliance on 'what everyone knows' becomes a relic of the past, such 'monsters' are as much grist for the mill of an inductive science based on singular facts as the comfortably familiar and typical quotidian phenomena readily available to all.

The previous chapter has already alluded to the manifestations of civility and modesty in the 'literary technology' Boyle laboured to create in support of his scientific claims. For here, as on the continent, experiment and documentation are closely linked. A specific literary technology had to be devised: 'a technology of trust and assurance that the things had been done and done in the way claimed' (Shapin and Schaffer 1985, 60). Boyle realized that if 'one wrote experimental reports in the correct way, the reader could take on trust that these things happened. Further, it would be as if that reader had been present at the proceedings. He would be recruited as a witness and be put in a position where he could evaluate experimental phenomena as matters of fact' (62–3). This took the form of a style of writing that presented the experiment through an 'ornate sentence structure, with appositive clauses piled on top of each other,' in order 'to convey circumstantial details and to give the impression of verisimilitude' (63). An ornate, rather than succinct, style was required to present simultaneously, in one snapshot, as it were, all the details required for virtual witnessing. 'Elaborate sentences, with circumstantial details encompassed within the confines of one grammatical entity, might mimic that immediacy and simultaneity of experience afforded by pictorial representations' (64).

Boyle's literary technology and his social technology of creating a community of 'modest witnesses' whose credibility can be successfully transferred to scientific claims are closely linked. The careful staging of scientific experiments played a critical role in his circle. Once the appropriate phenomena were generated in the laboratory, they were reproduced before a highly select group of credible witnesses. Such demonstrations 'were a routine feature of the meetings of the Royal Society, and a *Register-Book* was provided for witnesses to testify their assent to experimental results' (Shapin 1996, 107). But such direct witnessing, although crucial in the constitution of matters of fact, was very limited as a way of propagating the new and highly disciplined form of experience that was to legitimate scientific assent.[6] Boyle's literary technology allowed witnesses to be multiplied through 'the production in a *reader's* mind of such an image of an experimental scene as obviates the necessity for either direct witness or replication' (Shapin and Schaffer 1985, 60). His documentary practices transform readers into participants in a literary performance of the experiment through the eyes of the credible witnesses assembled by his social technology.

Boyle's literary and social technologies share an artfulness that attests to the labour involved in the manufacture of credibility. The literary technology is meant to represent what happened in the *staged* experiment – the 'results'– not the local contingencies of actual laboratory practices leading up to it. His 'circumstantial style' was designed as the prose version of this staged experimental scene. It was a scene already purged of the opportunism, contingencies, situatedness, and context of the reasoning that produced the phenomenon now materializing before the eyes of selected, reliable witnesses. Shapin's distinction between 'trials' and 'shows' is important here. 'Trying' an experiment, Shapin writes, 'corresponds to research proper, getting the thing to work, possibly attended with uncertainty about what constitutes a working experiment' (1988b, 399–400). Trials were risky and tentative: 'The notion of an experimental trial ... carried with it a sense of indiscipline: the experimenter might not be fully in control of the scene. The thing might fail. It might fail for lack of technical competence on the part of the experimenter, or it might fail for want of theoretical resources required to display the phenomena as docile' (401). A 'show,' on the other hand, is a staging of a successful trial, a 'display to others of a working experiment' (400). Trials were never displayed in public, whereas shows were. Thus the 'weekly meetings of the Royal Society,' Shapin notes, 'required not trials but shows.' These performances of Boyle's experiments, staged by his technical assistant Robert Hooke 'for the society's deliberation, instruction, and entertainment' (401), closed the gap between experimental and literary performance.[7] A performance of a successful experiment, made possible only by many carefully honed 'trials,' brings the performance unfolding before the fellows of the Royal Society much closer to the deliberately artful account of it fashioned by Boyle's literary technology than to a straightforward description of what had happened in an individual 'trial.' In attending a 'show' of one of Boyle's experiments, the Fellows were, in a sense, witnesses to a *documented* experiment, something much closer to the 'virtuality' of the literary account than to the experiment's trials. The artfulness of the literary document allows it to faithfully represent a correspondingly artful experimental 'show.' Thus the *circumstantiality* of an experimental report designed for virtual witnessing is the documentary correlate of a *theatrical* strategy of objectivity. Relying once again on the language of information, we can say that nature's information emerges from the laboriously artful documentary practices of the early Royal Society as a singular historical event authenticated and certified as a *matter of fact* which did indeed occur as documented. In Restoration England, objectivity retreated from the absolute certainty of deductive demonstration to the probabilistic moral certainty of inductive claims supported by the granularity of atomistic, singular, experimental events staged in a new documentary culture.

Lorraine Daston's work on the literary techniques of fact construction in the first scientific journals shows how much documentary work had to be performed in order to wrest the singularity and granularity of facts from the imaginings of popular culture and inscribe them in the appropriate Baconian form of objectivity. She shows that scientific 'facts' of the early modern period were quite different from ours. They were 'strange facts' – oddities, anomalies, curiosities, marvels, prodigies, and monsters. They were precisely the kind of 'strange phenomena' that were 'culturally salient in the first decades of the seventeenth century at all levels of society from learned humanist to illiterate milkmaid' (1998, 25). In popular culture, they were 'strange phenomena ... saturated with significance; all were grist for the interpreter's mill, and were as often as not pressed into service as propaganda on one or another side of the raging religious controversies of the day' (25).

But to gain the cultural status of scientific facts, their significance had to be transformed. Their 'strangeness' had to be maintained for Baconian purposes in order to ground inductive generalizations rigorously derived from singular, theory-resistant 'facts of nature.' Yet their strangeness for natural philosophy could not be the kind that fired the imagination of popular culture, because facts drenched with significance and meaning carried with them precisely the kinds of generalizations properly Baconian natural philosophers sought to avoid in rejecting Aristotelian first principles. They had to be stripped of their popular-culture strangeness and take on a different sort of strangeness, one serving a different set of cultural practices: the objectifying labours of Baconian natural philosophy. As Daston puts it, they had to become 'the very prototype of the inert datum, defying not only theological but also theoretical exegesis' (25). Baconian facts were not selected for their cultural oddity and fascination, but to cancel any licence for interpretive excess. They were 'handpicked for their recalcitrance,' and intended as 'a standing reproach to all extant theories' (24).

Strange facts, Daston writes, 'began as signs par excellence and ended as stubbornly insignificant' (25). How did this happen? How could they resist their popular significations to become 'notoriously inert – "angular," "stubborn," and even "nasty" in their resistance to theory and conjecture' (21)? How could they come to exhibit a radical singularity that freed them not only from the interpretive web of the popular imagination, but also from Aristotelian structures of deductive theories and explanations?

The problem, Daston argues, belongs to language. But not language spoken just anywhere, rather, the language inscribed in scientific documents produced according to Baconian strictures. She identifies three literary strategies – dissociation, narrative, and description – found in the early period of the *Philosophical Transactions of the Royal Society of London* for describing the kind of strange facts required by Baconian science (26–37). The first refers to separating the 'vulgar,' who eagerly seize upon strange facts with amazement and fright, from the 'curious and ingenious,' who were distinguished primarily by social standing, cultural status, and the qualities of 'esoteric learning, a cultivated sensibility [and] connoisseurship' (30). The labels 'curious' and 'ingenious' were well-known honorifics, marks of special status, imported into the natural-philosophical descriptions of the strange facts that littered the early *Transactions*. She also points out that a 'curious' sensibility featured a proprietary gaze, one that viewed strange phenomena as a kind of intellectual property, quite appropriate for propertied gentlemen of social standing. The second strategy consisted in the development of a special narrative form and genre: hence the short article rather than the long, speculative treatise, and a reliance on lists and tables in the documentation of strange facts in the *Transactions*. 'Lists,' Daston writes, 'carried with them a view of the structure of scientific experience, an experience that came in small chunks rather than smooth continua, an experience better suited to short journal articles than long treatises.' In a list, 'one item was linked to the next one only by an arbitrary enumeration that could be permuted at will' (37), thus offering documentary support for the detachment of facts from theory. The third strategy was a style of writing that featured 'multiple analogies, fine-grained and circumstantial detail, and frank aesthetic pleasure' (26). Because strange facts were theory-independent, they had to be described in excruciating detail, 'for who could tell which detail would turn out to be significant?' These descriptions, 'the literary equivalent of the impossible accuracy of some of Albrecht Dürer's animal paintings' (31), demanded herculean feats of exaggerated attention. Daston's fine sensitivity to the documentary practices on exhibit in the early *Transactions* also captures the meticulous attention paid to the sensualism of the surfaces of things, and the great aesthetic pleasure expressed in their descriptions, which displaced 'vulgar' emotions of fascinated disgust and horror. Other techniques included the use of multiple analogies to describe the various parts of things, thereby creating what Daston describes as a 'chimera of the imagination' (35). Numbers were also commonly used, but for precision (for clarity, distinctness, and intelligibility) rather than accuracy (35–6). Taken together, these documentary strategies worked to inscribe strange facts of the Baconian kind: 'Strange facts were thus detached in their descriptive elements both from natural philosophical theories on the one hand and from religious interpretations on the other' (36).[8]

Daston's work on strange facts shows how specific, historical modes of documentation function to constitute the objective facts of Baconian science. But strange facts had a relatively short life span. When the Royal Society adopted Newton's physics, this 'amounted to the reform of the radical and philosophically fruitless "experimental philosophy" of the Restoration period' (Dear 1995, 210). Newton's reliance on geometrical constructions in presenting experimental results grafted the certainty of mathematical demonstration onto claims of a 'mechanics' that the experimental event is depicted as exemplifying. Thus Newton's reliance on the mathematical form of argument mobilized resources long familiar to readers who demanded that universal knowledge claims be *evident* in the way that reports of singular events – in canonical Aristotelian fashion – could never be. But Newton also incorporated the event experiment into a form of reasoning that went far beyond his continental predecessor's attempts to make experiments typical of how things happen in nature. In the 'physico-mathematical' reasoning of Newton, events in nature do not happen because of the inner causes embedded in the essences of things. Universality has a different form than in its Aristotelian manifestation. For Newton, it took the form of mathematical laws grounded in geometrical constructions that depicted singular experimental events. 'Hence Newton's version of "experimental philosophy" postulated the actual production of particular phenomena so as to allow the formation of a universal science from singulars; the trick lay in that final stage and was accomplished, to the extent that it could be accomplished, by framing the issues in terms of physico-mathematics' (242). Since evidence for claims of physico-mathematics was grounded in the necessity of mathematics, it was possible to derive a 'universal from a single experimental setup, properly considered' (242). By giving 'event experiments a philosophical respectability that they had formerly lacked,' Newton's physico-mathematics achieved a convergence between the 'moderns' – experimenters who eschewed Aristotelian causes and essences – and the 'evidentness' of universal claims long required by scholastic cultural resources for making knowledge claims credible. Universality migrates from 'what everyone knows' about how things happen in nature (due to their essences) in continental Europe, flirts with the granular and strange atomic 'facts' of Restoration England's Baconian science and, with Newton's physico-mathematics, settles on the mathematical laws facts exemplify: 'The uses of mathematical argument in making accredited knowledge of nature show how the foundational assumptions of a mathematical science, and the kinds of experience that underwrote them, made it possible for Newton to announce a kind of declaration of independence for physico-mathematics' (210).

Bazerman's (1988b) study of the historical development of the *Transactions*, and especially of Newton's role in it, reveals the degrees of resistance presented by different documentary forms to alternative constructions of objectivity. Like Daston and Dear, Bazerman traces the evolution of the documentary articulation of scientific facticity, from the Baconian science of the early modern period to the theory-laden facts of Newtonian science, a transition from an isolated atheoretical singularity to embeddedness in a highly theoretical representation of the natural world. Bazerman's study maps this transition onto Newton's difficult negotiations between two documentary forms: the journal article and the book. Newton's difficulties with the former and comfort with the latter offer historical traces of the resistances and accommodations of different kinds of documents in their articulation of objectivity.

Bazerman reminds us that Newton made only one foray into the journal literature. As we have already seen from the work of Daston, Dear, and Shapin and Schaffer, the Baconian literary conventions of the early *Transactions* deliberately offered strong resistance to theoretical claims. To satisfy its readers (and its editor), Newton had to document his results according to the properly Baconian form of brute facts. 'His overall rhetorical problem,' Bazerman writes, 'is to give an account of his findings so that they appear as concrete fact, as real as an earthquake or ore found in Germany, even though the events that made these facts visible to Newton occurred in a private laboratory and as the result of speculative ponderings and active experimental manipulations' (90). The documentary form of description required was close at hand, because the 'personal account of stumbling across an unusual fact was a common one used in the early *Transactions*, as in the accounts in the first volume of the luminescent pickled mackerel and the putrefaction of maydew' (91). It was an enduring documentary phenomenon, because even until 1800, Bazerman notes, most of the *Transactions* 'were devoted to observations and reports of natural events, ranging from remarkable fetuses and earthquakes, through astronomical sightings, anatomical dissections, and microscopical observations' (65). In Newton's *Transactions* article, he 'mangles' his account to accommodate the prevailing form of documentary agency. Bazerman observes that 'Newton attempts to make his findings appear as concrete facts by establishing in a discovery narrative his own authority as a proper observer of concrete facts. This narrative presents him stumbling across a natural fact, as one would stumble across a rock' (90). The ingenuousness of Newton's notebooks, in which his theoretical motivations for his experiments are plain to see, were masked in his *Transactions* account – and when they were not, they risked the editorial excisions of Henry Oldenburg, the editor of the *Transactions*, who deleted from Newton's article a passage in which he expressed his certainty in 'the facticity of his generalizations' (98). Although utterly convinced of the robust facticity of his theoretical generalizations, Newton's account exhibited the 'curious and ingenious' sensibility of a Baconian experimenter, presenting himself 'as being moved by the phenomenon of colours itself and having an attitude of naive wonder at the spectacle of nature' and his observations 'as incidental to an interest in grinding nonspherical lenses.' Newton's account is the documentary performance of 'the Baconian collector, free of prior theoretical impulse, being only led into inquiry by the observed facts themselves' – someone who is not 'trying to find out anything in particular' until he stumbles upon a scientific curiosity (93).

Submission to documentary practices that constructed facticity in terms of a strangeness and granularity highly resistant to theoretical generalizations and explanations, and whose targeted Baconian audience was highly alert to and suspicious of any trace of hypotheses or theories surreptitiously introduced into presentations of concrete matters of fact, presented obstacles to the exercise of Newton's singular and revolutionary genius in documenting the natural world as a fully imagined, coherent, closed system operating according to universal, natural law. He never published in a journal again. Bazerman's account of Newton's replies to his critics documents the development of a new style and a new kind of account that became, in his books, what Bazerman calls a 'juggernaut' of persuasion. This new literary form consisted in 'integrating an actual experiment into a general geometrically styled argument' (115). It needed an alternative documentary form to the journal article as it existed in the early *Transactions*. The form of the book allowed Newton to lay out his *Optiks* 'in the manner of a Euclidean tract,' moving 'from definition to axiom to propositions' (119). It allowed for the mathematization of his entire system, and to embed empirical results in a structure of reason that offered massive resistance to the hesitations and doubts of Baconian 'modesty.' The first book of the *Opticks*, Bazerman writes, could elaborate a 'complete system ... presented as a logical and empirical juggernaut, with every step in the reasoning backed up with carefully described experiences precisely related to the formal proposition' (121). In his books, Newton developed fully and freely his critical step in the evolution of the literary technology of objectivity – from strange facts to physico-mathematics: 'The form of compelling argument he developed relied on creating a closed system of experience, perception, thought, and representation that reduced opposing arguments to error. The closed system Newton developed was his own, framed by the worlds represented in his powerful books. Only later was the science community to develop the means to construct communally developed closed systems; nonetheless, the Newtonian model of argument provided a powerful way of arguing for general truths from empirical experience' (83).

These stages in the history of universality show that it is, first, historically contingent and, second, as a cultural imperative it is a product of a complex set of practices, central among them documentary practices, that deploy a diverse set of cultural resources. None of these stages support an interpretation of scientific documentation as events in the history of the communication of information. It is not as if, for example, the information nature has to offer was simply more efficiently communicated as improved instruments were invented and observations were refined by experiment. That is not the role documentation played. Instead, natural-philosophical writing helped *construct* credible knowledge claims. Moreover, the role of writing in the history of universality shows that such constitutive effects of writing precede, and make possible, any communicative and informing effects. Problems can only be posed by the latter once the cultural resources of the former have been assumed, or, as is more usually the case, simply ignored.

Engineering Universality

Harding and David Hess provide many examples of contemporary non-universalist standpoints on nature, especially those pertaining to women's bodies, medical treatments, alternative therapies, environmental issues, and native knowledge systems.[9] Most of these cases occupy cultural niches far removed from the mainstream of legitimated, authorized, and institutionalized science. But contemporary practices offer other examples of legitimate kinds of knowledge production whose connections with universal knowledge claims are problematic and tenuous. Recent studies of technology suggest that engineering falls into this category. Insofar as theory plays a central role in the universalizing project, the degree to which engineering practices are pursued independently of scientific theory provides a measure of their autonomy from universal knowledge claims.

Engineering design, as Walter Vincenti puts it in his *What Engineers Know and How They Know It*, is a 'social activity directed at a practical set of goals intended to serve human beings in some direct way' (1990, 11). It provides an especially fertile field for studying the complex relationships between theory, experiment, and local practices that are focused on production of artifacts. Vincenti argues against the common view that engineering is merely applied science, in which 'engineers are seen as taking over their knowledge from scientists and, by some occasionally dramatic but probably intellectually uninteresting process, using this knowledge to fashion material artifacts. From this point of view, studying the epistemology of science should automatically subsume the knowledge content of engineering. Engineers know from experience that this view is untrue, and in recent decades historians of technology have produced narrative and analytical evidence in the same direction' (3). Research on the autonomy of engineering from theoretical science shows the cultural differences between these disciplines with respect to observance of imperatives to confer universality upon their results.

Thomas Hughes's history of ARPANET (the interactive, computer-based information network funded by the Advanced Research Projects

Agency of the U.S. Department of Defense), which later evolved into the Internet and the World Wide Web, is a good example of engineering design preceding theory. The design of interface message processors, the small gateway minicomputers that route message packets along alternate routes of the network, 'was essentially a problem in engineering design rather than the application of theory' (1998, 280). The team of engineers working on the problem, 'like the engineers and scientists who designed the heat shields for the Atlas and Titan missiles,' Hughes observes, 'could find little theory relevant for guiding their empirical thrusts, explaining their empirical successes, or rationalizing their empirical designs.' The remark of David Walden, one of the team responsible for software, that 'the academic analysis tended to come later' (281), expressed the view of most of the team, who, after the technology was developed, 'could argue that theory was playing its proper role of following upon and rationalizing practice' (191).

Another example is the design of the Davis wing. Although an unusual case, it illustrates vividly engineering's tolerance of distance from theory. David R. Davis was an entrepreneur and lone inventor who in 1937 designed the airfoil for the B-24 bomber, one of the most successful bombers of the Second World War (19,000 were built). The reasoning behind his design, as revealed in his notes, lacked any reference to theoretical knowledge of fluid mechanics. He used a geometrical procedure to generate a curve with 'an airfoil-like shape,' and then 'devised a complicated and unlikely geometrical construction to change this shape to something closer to a typical airfoil and translated this construction into equations by ordinary algebra and trigonometry. He gave no explanation of the reasoning behind his construction, which could not possibly have depended in any logical way on fluid mechanics' (Vincenti, 23). His experimental test of his airfoil's performance was equally inventive:

> Since no wind funnel was available, he improvised by borrowing a large Packard car from his friend Douglas Shearer, chief sound engineer at the Metro-Goldwyn-Mayer Studios and brother of the movie actress Norma Shearer. He then mounted a large, flat board horizontally on top of the car, to isolate his model from the aerodynamic disturbances of the car body, and tested his airfoils cantilevered vertically above the board. The measurements – of the distribution of pressure at the surface of the airfoil – were made by photographing an array of manometers in the car as it was driven at high speed (on lonely back roads in southern Califor-

nia, according to one source, and with flanged wheels on an abandoned railway track in the desert, according to another). (23)

Since Davis's experimental tests were designed to find the 'optimal airfoil among the family obtained by altering the values to the constants in his equations' (23–4), and since there were too many members of the family to test in his lifetime, Davis had to find another method to find the best one. In response to a question about it later, 'Davis said he ... "sat in a chair for three days considering the matter ... and concluded on theoretical grounds that plus and minus one were best." He did not say what the theoretical grounds were. He then checked this airfoil out to his satisfaction on the Packard. This was the airfoil incorporated in the model he delivered' (24). The results were 'so striking that when they were first obtained, it was felt that some experimental error must have occurred' (24). But the exceptional performance data held up, and the wing was used in the design and production of the B-24 bomber.

The Davis episode occurred during what Vincenti calls the decade (1930–40) of the 'adolescence of airfoil technology, when rational behavior was on the increase but offbeat things could still occur' (50). He is quick to point out that the 'Davis airfoil was not of itself very important for the history of aeronautics. It played only a questionable role in its one major application, and its contribution to airfoil technology was essentially nil' (50). Even the exceptional performance test data were of questionable significance, because the many factors that account for actual performance suggest that 'the Davis airfoil was probably no better than other contemporary airfoils would have been on the operational B-24, whose successful performance can be explained on other grounds' (49).

But it would be a mistake to assume that the uncertainties under which Davis laboured and the blindness of the design variations he made were *in principle* unique to his case. Although 'Davis's variations in shape were almost completely blind in any meaningful sense, virtually simple cut-and-try even though represented by sophisticated-looking equations' (48), Vincenti argues that the method of 'blind variation' – when 'blind' is understood to mean 'that variations take place, not randomly, but only without complete or adequate guidance' – is central to the knowledge practices of engineering. The main argument of his book is that the production and growth of engineering knowledge relies upon a 'variation-selection process.' Davis's 'almost completely blind' design variations were anomalous only in their degree of blindness, not in kind.

The air-propeller tests conducted between 1916 and 1926 by William F. Durand and Everett P. Lesley were also experiments on the effects of design variations. The tests were conducted, in contrast to those of Davis, by using rigorous experimental and mathematical techniques. Models of propellers varying in a specific set of design parameters were tested in wind tunnels, and the performance data were assessed according to known laws of similitude, which permit conclusions to be drawn about the performance of full-scale devices from test data on scale models. Their experimental parameter variation method, although conducted with a rigour far removed from Davis's ingenious pursuits, was nonetheless one which produced needed data in '*the absence of a useful quantitative theory*, that is, to get on with the engineering job when no accurate or convenient theoretical knowledge is available' (162; emphasis in original). 'Even in the 1940s,' Vincenti notes, 'airplane designers were still instructed, whenever suitable information was available, to select propellers on the basis of experimental parametric data' (156). Durand and Lesley's experiments provided just such data. The important point is that even though their work exhibited 'the gradual refinements of technique typical of experimental research: improved accuracy of measurement, more convenient presentation of data ... and growing sophistication of methods and ideas' (152), it had an important similarity to Davis's work: the use of a method – parameter variation – to bypass theory. According to Vincenti, this feature distinguishes engineering from science, and on that account is the most important of his findings regarding engineering's knowledge practices.

Vincenti is very clear about the need for further research to support his conclusions and to refine and deepen our understanding of what engineers know and how they know it. But his observations on the differences between science and engineering have implications for corresponding differences in how objectivity and universality are constructed – in both the documentary and non-documentary practices of these two different cultures of knowledge production. His research shows that theory is not absent from engineering, but is used, not as an end in itself, but as one of many tools to get the job done. Commenting on the struggle to establish a theory for propeller design, Vincenti notes: 'Engineers would have been happy to use the theory if the attempt had succeeded. When it failed to produce a theory of sufficient accuracy, however, they could still fall back on experimental parameter variation and laws of similitude to get on with the job of designing airplanes' (167). The method of experimental parameter variation itself, used by Durand and Leslie, 'can be carried out by either experimental or theoretical *means*' (160). For engineers, the value of a good theory consists in practical benefit: 'Even when an adequate theory of some sort is available, experimental parameter variation may still be employed because of lack of numerical data on the physical properties of the substances involved or insurmountable difficulties of one kind or another in carrying out the theoretical calculations. Whether theory is or is not available, the case for the use of experimental parameter variation often boils down in the end to the very basic one that it provides usable results in an acceptable time, whereas waiting for theoretical understanding or guidance may involve indefinite delay' (163).

The use of experimental parameter variation to *bypass* theory does not mean, however, that scientists do not use it, or that engineers do not use it for theory development. But engineers use it 'more for provision of design data than for the development of theory.' Since they are often required to produce results in the absence of theory, engineers depend upon the method more than scientists, 'whose paramount interest in theory can be served in other ways' (162). Vincenti's summary remarks on the use of experimental parameter variation by both engineers and scientists highlights the differences between engineering and science in terms of the importance of theory:

> In their common use of parameter variation in the quest for theory, however, engineers and scientists have different priorities. Engineers are after a theory they can use for practical calculations, perhaps eventually including the provision of parametric design data. To obtain such a theory they are willing, when necessary, to forgo generality and precision ... and to tolerate a considerable phenomenological component ... Scientists are more likely to be out to test a theoretical hypothesis ... or infer a theoretical model ... In either case their primary goal is to explain known phenomena or predict new ones. They value generality and precision for their own sake and consider it important that the theory or model be as close to first principles as possible. Essentially, scientists are interested in understanding, in some sense, how the physical world operates ... The distinctions are observable ones ... and are consistent with the view of scientists and engineers as communities with different values, the one valuing 'knowing' and the other valuing 'doing.' (161)

The importance Vincenti attributes to theory in the work of scientists is problematic, however, because studies of practices have shown that much of science is not pursued in the interests of theory, and that much scientific 'tinkering' looks very much like engineering practice – a point confirmed by Vincenti's observation of the use of experimental parameter variation by scientists. Moreover, his appeal to communities, values, and interests runs afoul of research that has failed to find the shared values and 'interests' often presumed to be involved in understanding 'how the physical world operates.' Such values, as previously argued, are demonstrable, if at all, more in what scientists say than in what they do; they surface in the empirical discursive repertoires of scientists, as Gilbert and Mulkay have shown (1984). Previous chapters have suggested that scientific documentary practices articulate a shared narrative field as an objective world of facts and universal knowledge claims, especially through literary techniques of occluding locality and substituting an abstract information space, thereby contributing to the labour of extending local research resources to collaborative, ongoing work. The key difference between scientists and engineers would appear, therefore, to derive more from cultural imperatives of their documentary practices than from different allegiances to theory. Studies of scientific practices reveal scientists as practical reasoners who make opportunistic resource selections in local contexts. Some are theorists, others are instrumentalists or experimenters. What distinguishes them from engineers are institutionalized documentary practices and corresponding literary technologies that articulate local 'contingency space' as abstract 'information space.' Engineers do not operate under as strict an imperative as scientists to 'mangle' their results according to strictures of objectivity and universality, nor are they under as strong an obligation to document their results in literary forms serving the universalizing project of research science. For engineers, it is enough to get the artifact to work. If theory needs to be developed to do that (given the available time), then theory will be developed. But engineers and scientists both face locality and contingency. The engineering job that needs doing does not present the special problem faced by scientists, of constructing universality from them. Scientists and engineers construct credibility in different ways. A research task for documentation is to show how these differences play out in their formal writing.

Towards Documentation

Studies of scientific practices analyse the authorized, legitimated forms of Northern/Western natural sciences as a disunified set of historically contingent knowledge practices. These practices not only fail to exhibit the consilience advocated by Edward O. Wilson (1998), but require concepts such as Peter Galison's 'trading zones' (1997, chapter 9; 1999), and Susan Leigh Star and James Griesemer's 'boundary objects' (1989) to make sense of their tentative and fragile relationships. It is a historically contingent feature of the cultural organization of scientific practices that those which articulate their results as universal knowledge claims enjoy a more powerful legitimation and exercise vastly greater authority than those which reject or are otherwise not engaged in the universalizing project.

Examples of alternative cultures of prediction and control of natural phenomena show that science need not be produced according to cultural imperatives of universality. One of the difficulties in discussing local knowledge systems is that universality is often used as a criterion for calling a particular activity 'scientific.' According to this usage, it becomes merely tautological to claim that any activity that produces stable phenomena or successful predictions and control over natural processes, yet which is not articulated in terms of universal laws of nature, is not scientific. Attempts to connect universality to science by definition are part of the universalizing project, not a response to empirical evidence that non-European knowledge systems are less effective than modern sciences. Harding puts it this way:

> No matter how effective other cultures' (including 'women's cultures') knowledge traditions are, were, or might have been for enabling effective interaction with natural worlds, they are not counted as real sciences. No matter how much modern sciences might have incorporated elements of other cultures' concepts and theories about nature, their mathematical and empirical techniques, and even whole bodies of their accumulated navigational, medical, pharmacological, climatological, agricultural, manufacturing, or other effective knowledge enabling prediction, and control of nature, these other bodies of knowledge are not counted as 'real science' until incorporated into European knowledge systems. And no matter how poor at explanation, prediction, and control European sciences are – for example, with respect to social causes of environmental destruction, or the causes of patterns of carcinogens or contagious diseases – these inadequacies do not count against European sciences' purportedly unique universal validity.' (1998, 166)

No merely verbal sleight of hand, however, can make universality a necessary feature of scientific knowledge.

When science is produced according to the universalizing imperative, the modern institutional form of its formal literature is central to it. When it is not, other documentary forms become relevant. Yet they confront serious obstacles in a culture of objectivity. Documentation of alternative knowledge systems faces the problem of institutionalized restrictions on what counts as a legitimate scientific document. This is not a problem of the communication of 'information.' The issue is more fundamental: the exclusion operates at a *constitutive* level, where it is determined how a document becomes informing in the first place – to whom, in which circumstances, and in which documentary circuits. Alternative knowledge systems face well-documented problems of universality's fellow travellers: colonialism, racism, sexism, religious intolerance, commodification, and imperialism. As a component of the universalizing project, the institutional form of science's literary technology is thoroughly interwoven with them. Seeing and responding to these exclusions is just one of the challenges for studies of scientific documentation, but one that first requires: (1) a shift of attention from the informational and communicative effects of documents to their constitutive effects, and (2) an acknowledgment of alternative systems for the documentary articulation of non-universalizing standpoints on nature. To view the journal article as a vehicle for the communication of information is already to implicate it in the representationalist and universalizing projects of modern science, and hence in their associated cultural alliances. It becomes something about which the question of its value as a vehicle for the communication of information makes sense, and which can become the primary analytical issue for information studies, only when the more fundamental, constitutive issues are occluded. An information science built upon the pillars of information and communication as theoretical kinds, rather than as cultural phenomena of specific historical moments is, in its application to science studies, part and parcel of modern science's universalizing project. It remains blind to the institutionalized documentary practices by virtue of which the phenomenon of information appears at all and to the circuits of economic, political, and cultural power that sustain them. Its conceptual, cognitivist, and representationalist biases prevent information science from taking us very far beyond a narrow range of issues about performativity of information retrieval technologies or design of systems and services in aid of more satisfying 'information behaviours.' Information is a minor player in materialist readings of scientific knowledge systems. Deflating information has implications for studies not only of scientific practices, but also of documentary practices generally. This section offers some resources for pursuing these implications, and a summary and overview of several issues that have been addressed from other perspectives in previous chapters.

Anti-Representationalism

Critiques of representation have long been familiar features of contemporary thought. Alternatives to a theory-dominated, representationalist interpretation of science has been a major theme of this book. Larry Grossberg's study of another cultural phenomenon – rock music – is a useful guide to the benefits of anti-representationalism for documentation. He starts from a critique of the model of culture as communication, which he depicts as follows:

> the model of communication assumes a relationship between two discrete and independently existing entities: whether between individuals, or between audiences and texts, or between signifieds and signifiers. The result is that any cultural relation takes on the form of an unspecified and unspecifiable exchange – a mediation – between encoding and decoding. It makes little difference which term in the model is given priority as long as the distance or gap between them remains. Whether the text or the audience ... has the power to determine the meaning of a specific communicative event, communication is the process by which that gap is overcome, the unknown becomes known, the strange becomes familiar. The model remains the same even if one hypothesizes a process of negotiation in which each is granted some power. The distance between them remains sacrosanct. (1992, 38–9)

The 'communication model' described here has obvious affinities with IC models. It also echoes several leading ideas familiar to readers of the LIS literature: the Shannon and Weaver metaphor of communication (or 'information transfer') as encoding and decoding, the belief that a shift to the individual user as the arbiter of meaning represents a significant step forward in our understanding of information use, and the notion of a 'gap' bridged by the user's acquisition of information. The effect of the communication model is a significant constriction of a broad field of inquiry. As Grossberg puts it, it signals 'the reduction of culture to texts and of human reality to the plane of meaning. The analysis of culture then involves the interpretation of cognitive, semantic or narrative content which lies hidden within the text' (43).[10] Proposals to index the 'content' of rock music by placing warning labels in its packaging are compelling examples of commitments to 'content which lies hidden within the text.'[11] From the perspective of IC models, such proposals privilege the 'messages' thought to be conveyed by its lyrics or musical style. By its representationalist emphasis on content as an independently existing entity – in the terms of this book, on 'information' as a theoretical kind – the communication model 'flattens the possibilities of cultural relations and effects' (44). Yet rock music, as Grossberg shows, is a cultural phenomenon with multiple effects, beyond those reducible to messages, meanings, representations – or information. His comment about texts generally are especially applicable to those effects of rock music that cannot be reduced to the information it is presumed to communicate: 'The same text can be a source of narrative romance, emotional support, sexual fantasy, aesthetic pleasure, language acquisition, noise (to drown out other demands), identity formation or rebellion against various powers. The active engagement with texts is rarely determined exclusively by the interpretive content of meaning production' (44).

The benefit of anti-representationalism for documentation therefore lies in broadening the range of inquiry. When the emergence of information is recognized as but one effect of documentary practices, attention can turn to others. I have argued that the construction of scientific objectivity and universality is one of the significant effects of the institutionalized documentary practices of producing and deploying scientific journal articles. It is not the case, therefore, that communication of information is the only or even primary effect of documentary circuits and practices. Communication of information is a *representational effect* but there are others, as Grossberg insists: 'If not every meaning is a representation, and not every text has representational effects, it may also be true that texts may have effects other than meaning-effects, and meanings, interpretations, uses and pleasures may themselves have additional effects ... (whether one calls them meaning-effects is not so important as recognizing that they cannot be reduced to the level of cognitive and narrative systems). Cultural texts are always at least potentially multifunctional' (45).

The challenge for documentation is to acknowledge practices with discursive objects in which information and communication play only a marginal, if any, productive role. When research is configured by the highly restrictive notions of information seeking, information use, and information behaviour, severe limits are placed upon understanding the full range of cultural engagements with discursive objects. For example, the benefits and burdens of documentary practices among marginalized groups are of limited use if analysed exclusively in terms of seeking, encountering, communicating, or behaving with respect to 'information.' The role of documentary practices in producing and acting out alternatives for action, in constructing solidarity and alliances, or in enacting strategies of self-presentation and negotiation on an often dangerous cultural terrain frequently have little to do with seeking and using information. Studies of the role of documentary genres such as science fiction, horror, romance, fantasy, crime, and many more, in the production, maintenance, and eventual fading of the multifarious identities that populate various sites of popular culture are not studies of information seeking. Concepts of information seeking or use contribute little to analyses of documentary practices with romance novels or pornography, for example, yet the complexities of both genres have received serious investigation from perspectives far removed from information studies.[12] A final example of research that pays attention to non-representational effects of documents are those concerned with the materiality of documentary forms: the book, print, television, film, computer-based electronic communications, radio, the Internet and World Wide Web, and numerous others. The communication of information plays a minor role in understanding, for example, the effects of print on science and culture.[13] When attention turns to effects of documentary practices analysis can move beyond the messages – the 'information' – that documents are thought to convey.

Anti-Essentialism

Anti-essentialism is another familiar conceptual resource. Once again, Grossberg is a useful guide. He challenges an assumption even more basic than the communication model: 'a principle of interiority or essentialism which locates any practice in a structure of necessity and guarantees its effects even before it has been enacted.' According to such a principle, 'any event – e.g., a particular political activity, an economic relation, a social identity or even a cultural text – is assumed to already contain its own identity, and its place in a history of transformation can only involve spinning out the associations, relations and correspondences already inherent within that identity ... Everything seems to be sewn up, stitched into place, guaranteed in advance' (52–3). Rejecting this principle is related to insisting on contingency, singularity, and historicity. If events have no fixed interiors, no stable identities to account for their relationships, and no guarantees as to their outcomes, they are not known by appeal to universal law or logical relationships between fixed meanings. The study of their unfolding becomes historical: how a singular and specific event or set of events develop at particular times and places in relation to other events. A practice is defined, not by the law of its identity in accordance with which its history is determined, but by its contingent effects. 'If there is nothing essential about any practice,' Grossberg says, 'then it is only defined by its effects; it is in the production of its effects that the identity of a practice is given. To say that a practice is defined by its effects is to locate the practice in its connection to its exterior, to that which is other to it ... Paradoxically, a practice is not where it is (enacted, for example) but at all of those sites where its existence makes a difference in the world, at the sites of its effects.' Moreover, the production of effects depends upon contingent practices: 'The ability of a practice to produce specific effects, to produce this effect rather than that, is precisely what has to be constantly made and remade' (53).

Anti-representationalism and anti-essentialism are related. Because the identity of a text, on a representationalist account, consists in its meanings, it is therefore explicated in terms of something interior to it: an essence, or an intrinsic property. A representationalist view of texts is therefore an essentialism about meaning. According to this view, the effects of texts are guaranteed in advance by the logical and semantic relationships between their interiors – their meanings. Applied to scientific documentary practices, this view construes it as producing propositions and forging logical relationships between them through the application of scientific rationality to an intrinsic property: the epistemic content – the 'information' – of the statements expressing those propositions. By contrast, an anti-essentialist account of scientific documentation views the stability of its statements or other discursive elements not as something fixed ahead of time and which guarantees in advance their mutual relationships, but as an effect of the practices in which they and many other heterogeneous elements are engaged to manufacture scientific phenomena. Anti-essentialism replaces the idea that the relationships between discursive elements are determined by their interiors, or essences – whether these be thought of in terms of propositional or epistemic content, meaning, mental representation, or information – by historical accounts of contingent relationships forged between them. Documentation investigates discursive elements wholly exterior to each other; no intrinsic features can account for either the relationships between them or between them and other elements of scientific work. A documentary study of the scientific journal article describes the labour involved in constantly making and remaking contingent connections between the articles' statements and between them and the other elements of scientific work in the manufacture of scientific phenomena.

In his description of cultural studies of scientific knowledge, Rouse explicitly acknowledges anti-essentialism in science studies. 'Cultural studies of science,' he writes, 'reject the existence of an essence of science or a single essential aim to which all genuinely scientific work must aspire' (1996, 242–3), and refuse 'to require distinctive methods or categories to understand scientific knowledge as opposed to other cultural formations' (239). His insistence on the disunities of the sciences supports his deflationary account of scientific knowledge, that is, his dissent from the view that such knowledge 'belongs to a single kind similar or distinguishable *in kind* in any interesting way from other kinds' (243–4). The argument of chapter 3 extended Rouse's deflationary account of knowledge to information and communication. Studies of scientific documentation recognize that, like knowledge, scientific information is not a single kind of thing. It is, therefore, not a kind of thing inherent in a text as its meaning or epistemic content, nor a kind of thing present to the mind of a reader or 'information user.' Nor is it the kind of thing that provides the underlying principle of a presumed unity or coherence of behaviours of 'information users' – their 'information behaviours.' Scientific texts become informing, not by containing a particular kind of thing – 'scientific information' – which scientists are thought to seek, but by the culturally contingent practices – no different in kind from other cultural practices – whereby particular statements gain stability by figuring as resources in further practices. Documentation investigates both informative and non-informative effects of documentary practices by attending to their historical and cultural contingencies.

Analysis of systems of scientific documentation gain a resource in anti-essentialism because an abstract, theoretical kind – information – is replaced by material, discursive objects (i.e., documents): occasioned utterances, inscriptions, graphic representations, and so on. Idealized interpretations of relationships between texts are replaced by studies of documentary circuits as more-or-less stable products of contingent, opportunistic, highly situated, and material practices by means of which documents become part of the epistemic packages deployed as resources for further scientific work. Institutionally disciplined documentary practices contribute to the 'mangle' of the scientific labour that renders contingent and local laboratory phenomena as unfolding according to the script of Nature. The source of the resistance and agency of scientific documents do not derive from logical relationships between the information they convey – the epistemic content of their statements – but from labour processes through which the documentary imperative is made and remade. Anti-essentialism in documentation therefore directs attention to processes whereby contingent connections are forged between documents wholly exterior to one another, and between them and other elements of the practices in which they are engaged.

Articulation

A third important resource for the analysis of documentary systems is the concept of articulation. Grossberg says that his 'revised cultural studies' has two main commitments: 'a return to the real via a materialist theory of effectivity; and a principle of contextuality understood through the practice of articulation' (1992, 46). He explains articulation as follows: 'Articulation is the construction of one set of relations out of another; it often involves delinking or disarticulating connections to link or rearticulate others. Articulation is a continuous struggle to reposition practices within a shifting field of forces, to redefine the possibilities of life by redefining the field of relations – the context – within which a practice is located' (54). Articulation therefore refers to links between practices, to their mutual accommodations and resistances, and to the effects of their interactions. Rock music once again provides a clarifying example. Practices with popular music are linked to making and remaking the social distinctions of youth culture. The music, its videos, packaging, advertising, concerts, associated fashions in clothing, styles of personal presentation, and consumption of related products in other regions of the cultural field, such as film, magazines, websites, and even food – in short, the cultural formation of popular music – are connected to practices of social distinction that differentiate youth according to musical categories: heavy metal, acid house, country, soft rock, folk, swing, and many more. These terms come to stand not only for genres of and preferences for popular music, but are used as indicators of social positions in the broader field of youth culture. Practices with popular music are therefore *articulated to* practices of social distinction. One set of relations is constructed out of another. As many fans of heavy metal music and their parents can attest, 'articulation is a continuous struggle to reposition practices within a shifting field of forces, to redefine the possibilities of life' (54). And if the guardians of traditional values could break the spell of their fixed gaze on the 'messages' or 'information' they are convinced are sewn up in the interiors of rock music, their nervousness might evidence some recognition that rock music's articulation to social practices 'often involves delinking or disarticulating connections to link or rearticulate others' (54).

A second example comes from feminist studies, and feminist science studies in particular. Scholars such as Evelyn Fox Keller, Susan Merchant, Donna Haraway, Sandra Harding, and Barbara Duden have shown how gender distinctions are articulated to scientific practices. Resistance to such studies often takes the form of appeals to the meaning or epistemic content of scientific statements – the 'information' they 'communicate' – which, so the argument goes, cannot send a gendered message because it is *purely* a representation of the natural, not social world. By abandoning notions of the interiors of statements, feminist cultural studies of science reveal that many connections forged between them are highly contingent effects of articulations of one set of practices – that which continually makes and remakes gendered social relations – onto another, that which manufactures scientific knowledge.

The concept of articulation is also central to Rouse's conception of cultural studies of science. He asks: 'So what are *cultural studies of science*? I use the term broadly to include various investigations of the practices through which scientific understanding is articulated and maintained in specific cultural contexts and translated and extended into new contexts. "*Culture*" is deliberately chosen both for its heterogeneity (it can include "material culture" as well as social practices, linguistic traditions, or the constitution of identities, communities, and solidarities) and for its connotations of structures or fields of meaning' (1996, 238). Rouse has drawn attention to the many articulations of scientific practices. Some involve extending laboratory practices to other scientific research contexts; others extend scientific practices to trans-scientific fields, especially those colonized by government, military, and corporate, commercial practices.

Until quite recently, Rouse and other students of scientific practices have paid little attention to the documentary articulations that I argue are central to the cultural formation of modern Northern/Western science. And when documentary practices are recognized, they are sometimes insufficiently materialized, as in the case of Pickering's (1995) treatment of the role of conceptual systems in the mangle of practice. His talk about concepts and conceptual systems in terms of disciplinary agency is a step in the right direction, because it tries to avoid metaphysical notions of conceptual agency. But conceptual agency operates through specific institutions, procedures, and technologies, among them those involved in the production and use of documents. The agency and resistance of conceptual systems can be at least partly accounted for by their materialization through institutionalized practices of writing, especially those governing science's formal literature. I have argued in chapter 4 that Foucault's (1979) concept of the institutionalized materiality of a statement leads directly to replacing talk of conceptual systems and their role in the mangle of practice with studies of the agency and resistance exerted by continually making and remaking the articulations of documentary practices – the disciplined operation of what Shapin and Schaffer (1985) call the literary technology of science – to the material, technological, and social practices of science. It is precisely because of the intertwining of the non-documentary practices of science with language, symbols, drawings, shop talk, ideologies, and the production of documents of all kinds that the entire ensemble is properly called a discursive formation, or a discourse, in the Foucauldian sense.

Although the distinction between documentary and non-documentary practices is artificial, a distinction drawn for analytical purposes only, the concept of articulation is a resource for understanding how the material, social, and technological practices of the natural sciences are related to their documentary practices. I do not wish to claim that there is a set of purely material, technological, or social practices that exist beyond documentation, and which only later get 'written up,' 'talked about,' or otherwise depicted, like clothing a naked body. The aim instead is to see how practices are intertwined, how a 'hybrid' phenomenon, a 'quasi-object,' at once material, social, technological, and documentary, emerges from them.[14] Yet the intertwining of practices does not prevent our drawing useful analytical distinctions between the agencies, resistances, and imperatives of matter, society, technique, and documentation. Studies of scientific practices have shown that the documentary practices of science's formal literature, those governing the writing of the journal article, are not representational practices. They are not linked to laboratory processes and phenomena by virtue of representing them but instead by re-inscribing them in a de-localized narrative of objectivity. The concept of articulation draws attention to the mutual accommodations and resistances arising from the interactions between these documentary practices and the others involved in laboratory life. It also draws attention to the highly contingent effects of articulating laboratory practices to the documentary practices of science's formal literature. I have argued that one such effect is to implicate the documentary construction of objectivity in science's universalizing project.

Inside Out

The concept of articulation presses issues and problems surrounding the relationships between documentary and non-documentary practices.[15] Articulation thus leads directly to a fourth resource for studies of documentary practices: their cultural openness. Since articulation refers to contingent interactions between sets of practices, there is no a priori delimitation of the range of practices to which specific practices can be articulated. With respect to scientific practices Rouse notes: 'The sciences are taken to be cultural formations that must be understood through a detailed examination of the resources on which their articulation draws, the situations to which they respond, and the ways they transform those situations and have an impact on others' (1996, 239). The sciences respond to and have an impact on a wide range of situations, which cannot be guaranteed in advance. They are not limited to the highly circumscribed range of conceptual situations thought to arise from the workings of an idealized scientific rationality, such as Popper's (1965) conjecture and refutation. Nor are they circumscribed by a unique representational objective, a distinctive method, a particular community based upon shared norms or paradigms, or a shared vocabulary or language. Since scientific practices are located in an indeterminate field of practices, there is in principle no distinction between those 'internal' to science and those 'external' to it, which are often construed to be related to the former in terms of 'context,' 'influence,' or 'diffusion.' 'By contrast,' Rouse explains, 'cultural studies of science take as their object of investigation the traffic between scientific inquiry and those cultural practices and formations that philosophers of science have often regarded as "external" to knowledge' (1996, 239). 'Science,' observes Steve Fuller, 'is one of the most loosely structured social practices' (1992, 421).

Cultural openness is a resource for studies of scientific documentation. Reliance upon concepts of information and its communication support internalist accounts of science. When the question of interest for the journal article is posed as whether or not it communicates the information required to create scientific knowledge, it is construed in terms of its contribution to a particular, bounded representational practice, one governed by a unique method internal to scientific knowledge production. But when the documentary practices of science are recognized as culturally open and thoroughly 'mangled' with others, new avenues of inquiry are revealed. Once the barriers between internal and external practices are broken down, investigations of traffic in scientific documentation across cultural fields shed the stigma of violating the true, inner heart of scientific knowledge production. Alignments between documentary and laboratory practices are in principle no different than alignments between documentary practices and many other practices beyond the laboratory. To pursue any scientific practice beyond the laboratory is not to venture beyond the borders of science to its 'social context,' but instead involves mapping traces of its cultural extensions.

There are many examples of studies that document the cultural openness of scientific practices. Galison's study of Einstein's and Poincaré's relativity physics situates both in 'a story in which physics, engineering, philosophy, colonialism and commerce collided' (2003, 41), a story in which there are no fixed borders between what is internal and external to science. Andrew Ross's (1991a) study of New Age science reveals a cultural openness that permits curious intertwinings of authorized research science with popular culture. Especially pertinent in terms of documenting mutations in forms of universality and objectivity is contemporary science's openness to corporate and military practices. And in her analyses of the role of universality, Harding has argued that there are few stronger allies of the Euro-centric masculinity of modern science than nature itself (1998, 164–5).

Hess's observations on the narrowing gap between cultures of academic and non-academic scientists draws upon science's cultural openness to exercises of power and domination. 'In the United States,' he writes in a comment increasingly applicable to universities in neoliberal economies, 'the old vision of a university as a community of scholars and scientists organized along democratic principles is increasingly cast aside in the race to the technoversity, in which profits come from high-tech ventures such as biotechnology institutes ... As the business/university relationship tightens, TQE (total quality education), restructuring, reengineering, downsizing, and other new managerial fads and techniques become part of the administrative culture, and departments are forced to justify their existence and budgets in economic terms' (1995, 166). Hess's 'old vision' of a democratically organised community of scholars should not be interpreted as a representation of a bygone social order. Instead, it is a discursive construct that deploys disinterested search for truth as a legitimating mechanism to graft a democratic social order onto universal science. But in a global culture characterized by the enormous power of the rhetoric of market forces as the only 'effective' and therefore legitimate principles of social organization, a democratic rhetoric of disinterestedness in the pursuit of universal truth becomes merely nostalgic, even somewhat quaint. However the current manifestation of market incursions into academia play out, the threat they pose to the objectifying strategies of contemporary science provides both another proof of the cultural make-up of the universalizing project and the problems attendant upon the articulation of science to commerce.[16] Nature and commerce cohabit the cultural landscape of twenty-first-century market economies in an uneasy tension, since on the one hand nature is deployed to legitimate commercial science (consider the corporate use of science in the tobacco wars and global warming debates) but, on the other, corporate imperatives of scientific activity undermine the autonomy of scientific élites which the universalizing project has continued to underwrite. Its historical fortunes under market conditions of scientific production do not tell a tale of the battle of reason or nature against commerce, but provide a map of shifts in the cultural conditions for making things work, how conflicts about what things are worth making are closed, who makes such decisions, and how the things made are distributed and used. The shifts in the cultural openness of the academic research laboratory to the corporate laboratory, from academic research projects to corporate projects, and from laboratories to diverse ventures of transscientific fields map the shifts in the discourse of universality, its institutional platforms of enunciation, its effects, and the changing contexts in which it is deployed.

Science's cultural openness suggests that a project for studies of scientific documentation is to chart through their documentary traces the trajectories of articulations of scientific practices to others. As the discussion of different forms of objectivity in science and engineering has suggested, this is not a project of mapping channels through which information flows, but of tracing various documentary constructions of credibility of different kinds of discursive objects. For documentation generally, the moral of cultural openness is that documentary practices are not bounded by internal or a priori properties that determine their articulation to other cultural practices. Cultural openness is itself the product of practices, some much more powerful than others.

Reflexivity

A fifth resource for analyses of documentary systems is reflexivity. Investigations of documentary practices are, like the practices they investigate, culturally situated. Grossberg's characterisation of cultural studies also applies to documentation: it is 'always a strategic intellectual practice' (18). Rouse says much the same thing by noting the dissolution of 'the boundaries between cultural studies of science and the scientific practices they study ... Cultural studies' interpretive readings are thus part of the culture of science and not an explanation or interpretation of it from "outside"' (1996, 255). Just as cultural studies of science recognize 'that claims to knowledge are invariably entangled in relations of domination and empowerment' (254), they are also sensitive to the entanglements of their own practices. Contrasting cultural studies of science to a 'tradition that posits value freedom as a scientific ideal,' Rouse notes that they 'have a stronger reflexive sense of their own cultural and political engagement and typically do not eschew epistemic or political criticism ... One cannot not be politically and epistemically engaged' (257).

Reflexivity is a resource for documentation because it directs attention to its own political engagements. In her discussion of universality claims that she argues are 'epistemologically, scientifically, and politically dysfunctional,' Harding notes that such claims state that the sciences 'simply provide pure information about nature's order' (1998, 169). This 'pure,' politically neutral information is what IC models present as the substance processed by the complex machinery of information technologies and information systems and manipulated by the quasi-scientific procedures of information and knowledge management. An information science that constructs information and communication as theoretical kinds is itself a politically entangled discursive practice. Its occlusion of political, historical, and cultural situatedness provides another resource for the politics of representation, the disciplines of conceptual clarity, and universality.

Rouse points out that cultural studies of science 'are located within ongoing conflicts over knowledge, power, identity, and possibilities for action.' They are 'reflexive attempts to strengthen, transform, or reconstitute existing alignments or counteralignments by situating them historically or geographically' (258). Turning the reflexive lens on this book, I acknowledge my intention to step into what I see as the disciplinary politics of information studies. My arguments against IC models and for documentation are aimed at challenging and, hopefully, making a contribution to breaking the spell cast over information studies by epistemological concepts of information and communication. The argument that the discipline's research field is enriched by studies that do not privilege information or communication as theoretical kinds, that do not depend upon IC models, that decentre epistemological presuppositions and concerns, and that attend to the many and varied practices with documents, to their systems of production, organization, and consumption, their materiality, the social disciplines required to maintain them, and their histories, is meant as an intervention in a disciplinary focus on issues of efficiency, management, and technological performativity at the expense of questions that can deepen our understanding of how documentary labour is intertwined with a wide variety of cultural practices.

The Future of Information Studies

Since a rejection of IC models involves the rejection of the concept of information as a theoretical kind, it would appear to undermine the possibility of information studies. It is therefore time to press the question, if information and communication are decentred and replaced by conceptions of practices with discursive objects – occasioned utterances, inscriptions, graphic representations, and so on (i.e., documents) – are information studies even possible? Does information not dissolve as an analytical category? Or are there compelling imperatives to retain it?

There are different approaches to these questions. One is strategic. To retain uses of the term 'information' in a distinct field of study may be politic, due to the historically contingent, disciplinary entrenchment of a term that may still function to differentiate a (loose) set of research, pedagogical, and professional practices from others on the academic and scholarly scene. Retaining the term as an emblem of documentation may help stabilize its study in an existing institutional site – a strategy akin to branding. The shift in focus advocated in this book – the study of a wide variety of practices with a wide variety of discursive objects – might then be achieved by nominalist strategies. Studies deploying a nominalist interpretation of information can retain commitments to historicity and contingency by hospitality to the historical origins of current pedagogical, professional and research practices and to librarianship's late-nineteenth- and early-twentieth-century engagements with documentation and its subsequent historical encounters with information science, information management and now, knowledge management, competitive intelligence, and their disciplinary fellow travellers. This catholic configuration of information studies, suitably historicized and acculturated, would also embrace issues of performativity and efficiency arising from the historical origins of information science. Although its uses of 'information' would be regulated only by a rejection of the notion that information studies relies upon a predetermined essence of information as a theoretically coherent kind, it would nonetheless welcome attempts to define information and communication as theoretical kinds, since such theoretical projects belong to a particular set of disciplinary practices with histories of their own. Such nominalism enjoys the advantages not only of catholicism, but also acknowledges explicitly that in current information studies, 'information' is used as a 'portmanteau' word, used merely to point towards an open-ended series of instances disciplined not by a conception of information as a theoretical kind but by the contingent, disciplinary history of information studies and its projection into the future. Reflexivity would compel a recognition that the imperative to extend information studies beyond the narrow focus of IC models itself arises only from historical contingencies that elevated 'information' to a leading trope of contemporary culture.[17]

There is, however, another, less catholic but equally political answer to questions about the possibility of information studies. Given the deflationary account of information provided in chapter 3 and invocations of IC models simply by using 'information' as a grammatical substantive, a better alternative may be to carve out a special field of study with more specific aims and fewer disciplinary encampments. As a grammatical substantive, which even 'portmanteau' usages cannot avoid, 'information' appears as if it were a theoretically coherent kind. Its grammatical form carries with it a penumbra of mentalism, knowledge as representation, and information as epistemic content – all the grammatical baggage and totemic concepts by virtue of which theoretical significance is breathed into studies of information retrieval, information seeking, information use, and information behaviour. It is unlikely that rigorously nominalistic police actions can overcome the resistances offered by deployments of 'information' as a grammatical substantive to investigations of the ways in which practices with documents are articulated and the effects their articulations produce, in the face of the IC models such deployments carry along in their wake. It may be time, therefore, to discard information altogether as a useful analytical concept and focus instead (and once more) on documentation as a surer guide to an area of study that wishes to draw attention to the utility of methods and approaches like those used in the social studies of science for investigating the historicity, materiality, and institutional environment of the complex social practices that sustain documentary systems.[18] The questions of interest centre on how documentary practices and the circulation of documents of all kinds, together with their related technologies of production and consumption, fit into social and institutional structures and produce specific effects, among them but not limited to: objects of theoretical and disciplinary knowledge, cultural phenomena, social distinction, hierarchies of expertise, social relations and institutions, and political action. Documentation is distinguished from information studies by locating concerns for information in the broader context of *constructive and constitutive effects* of documentary practices. In addition to studies of scientific practices, such an area of study draws especially on political economy, cultural studies, current research on the materiality of documents (e.g., controversies about 'oral culture,' 'print culture,' and 'electronic culture'), theoretically oriented issues arising in document conservation and preservation work, studies of the history of the book and other documentary forms, critical theory focused on the concept of the document, and post-structuralist forms of discourse analysis. Like the social studies of science and technology, documentation pays close attention to the many heterogeneous elements – technologies, institutions, social discipline, and history – from whose interactions there emerges, *among other things*, information itself. The final chapter indicates some research directions.

7
Documenting Stability

The studies of scientific practices surveyed in this book have set a problem for the study of scientific documentation: how do the documentary practices of science intersect with the other practices of scientific culture in the production of scientific phenomena? Posed in another way the question is, how are the practices that stabilize inscriptions related to those that stabilize scientific phenomena and their related social routines? A more general form of the problem is how to trace the many and varied effects of documentary practices at various cultural sites. The previous chapter identified some analytical resources drawn from studies of scientific practices for use in documentary studies. The present chapter extends this generalization in two ways. First, some directions for future research are indicated, not only for scientific documentation but for documentation more generally. Second, the documentary effects of stabilizing scientific phenomena are extended from science's primary literature to the 'masses of documents' comprising its organizational apparatus.

Research Directions

Translation

The first task of a shift from information to documentation is to translate talk about ideas, concepts, and information into talk about occasioned utterances and inscriptions, and talk about conceptual processes involving ideas, concepts, and information into talk about public practices with occasioned utterances and inscriptions. The Wittgensteinian example of completing a simple arithmetical equation, used in chapter 4, makes the point: the correct completion of the inscription '3 + 4 =' in our cultural practice of elementary arithmetic depends not upon what might be imagined as occurring in the 'mind' of the reader of this inscription, but instead upon what gets uttered, written on a page, typed on a keyboard, or clicked in columns of answers on interactive web sites.[1] If readers write '7' (understanding by 'writing' doing any of these things), their completion of the inscription is correct, no matter what images or representations may have floated before their minds (be it the conscious mind or, in such instances, the more popular, unconscious one). What goes on in minds, what might be present to consciousness upon 'grasping the meaning' of the inscription's elements, and what 'information' might be mentally processed to arrive at the result: none of these alleged occurrences are part of our criterion for saying that '7' is the correct completion. What matters is which inscription is produced in appropriate circumstances. In its investigations of what Grossberg calls the 'meaning-effect' – which might also be called the 'information effect' – documentation focuses on a document's inscriptions, not upon the abstract or conceptual entities that they might be thought to represent. At least four avenues of investigation emerge from the translation of the idiom of representation into the idiom of material practices:

1 The ways in which inscriptions contribute to the stabilization of networks of heterogeneous elements to produce particular phenomena, whether in a scientific laboratory or elsewhere: how do cultural imperatives to take specific documentary forms (such as, in science, the journal article) *into account* help construct such networks?
2 The contribution of the inscriptions of specific documentary forms to resource extensions: Rouse's work suggests investigations of the contribution of inscriptions to the narrative structure of cultural activities;
3 the ways in which inscriptions themselves are stabilized by virtue of their relationships to other members of the networks of heterogeneous elements constituting stable phenomena; and
4 the many different ways documents become integrated into various cultural practices, that is, mapping documentary practices as resources for the production of different sorts of effects.

Inscriptions as Actants

The second component of a research program for documentation consists in charting both the agency and resistances of inscriptions. Hacking's comment nicely captures what is at stake in such rethinking: 'Maybe I go further than Latour, for I might take inscriptions to be among the actants, right up there with fishers and molluscs, working and worked on, everywhere people go since the moment that our species came into being as *Homo depictor*' (1992, 36, note 1). The discussion in chapter 4 of Morpurgo's hunt for the free quark demonstrates how the inscriptions of science's formal literature exhibit agency and resistance. That discussion translated Pickering's claim that 'the dialectics of resistance and accommodation can arise in conceptual practice' (1995, 117) into a claim about the resistance and accommodation of inscriptions. Morpurgo's problem, as we saw, was not that his sentence, 'electrical charge is continuously divisible,' fails to represent the world truly – it represents what he found in his laboratory – but that he is working in a cultural field in which this inscription could not survive on the platforms of enunciation upon which scientific statements are proffered as scientific fact. There are too many other stable inscriptions massed against it. The interpretation of inscriptions as actants opens onto the following research areas in documentation:

1 charting the range of effectivity of agency and resistance, since there is no a priori reason to believe that they are uniformly distributed across the entirety of particular cultural fields (such as natural science), rather than being limited to particular regions whose specification, enumeration, and analysis become research problems;
2 mapping differences in the operations of resistance and agency of the same statement, set of statements, or documents in different regions of a cultural field; an inscription may stabilize and gain strength in one area, enabling it to offer strong agency or resistance to material, social, and discursive arrangements by virtue of its strong ties with other network elements, but exhibit very weak resistance elsewhere;
3 mapping the temporality of resistance and agency; some stabilizations of inscriptions last for a long time, and others are short-lived; how are these differences distributed over different categories of documentation, such as theory, technique, and instrumentation?
4 identifying the source of an inscription's power of resistance: since the alignments of heterogeneous elements that constitute a scientific phenomenon extend beyond the laboratory to include commercial, corporate, governmental, and military institutions, the sources of inscription's power of resistance are many and varied;
5 mapping the degrees to which the cultural practices of different sci-

ences depend upon inscriptions (Knorr-Cetina (1999) shows that sciences vary in the degree to which they engage inscriptions: high-energy physics, she argues, features 'a loss of the empirical' in favour of symbol manipulation).

Beyond Legitimation

The emphasis placed by IC models on the construction of conceptual fields as the goal of scientific activity results in a focus on institutionally authorized and legitimated science. By contrast, an emphasis on documentary practices expands the range of cultural sites, and hence the range of research problems, in which documentation contributes not only to knowledge production but also to stabilization of various other cultural phenomena. The following four points expand on this claim:

1 The translation of representations into inscriptions-as-actants means that inscriptions are not privileged in scientific knowledge production. When stability, not representation, is the goal of scientific activity, inscriptions and documents are but one of the many kinds of network elements that constitute scientific phenomena. Latour has shown how far stabilizing resources extend beyond the laboratory bench. Research into the contribution of an inscription, a set of inscriptions, or particular document forms to the stability of various cultural phenomena can lead to institutional sites far beyond those immediately apparent.
2 Materialist approaches to scientific practices challenge the unity of conceptual fields and its associated universality claims. Rather than a single science in pursuit of a theoretical representation of nature, many different and often unrelated sciences have diverse and varied purposes. Many local practices are held together in a fragile alliance, by resource-relationships, epistemic alignments, and epistemic things. Modern Northern/Western science turns out to be a gerry-mandered structure consisting of disparate networks of overlapping and intertwining threads. A legitimate research problem, therefore, is to identify the ways in which inscriptions and documents stitch together disparate fields of scientific practices. As indicated in the previous chapter, Galison's (1997, chapter 9; 1999) 'trading zones' and Star and Griesemer's (1989) 'boundary objects' are moves in this direction.
3 The turn from representations and information to documentation

helps reveal the precarious borders between legitimate and illegitimate, authorized and unauthorized science. Since unity and universality turn out to be, at best, projects of legitimation of a specific cultural form of knowledge acquisition, then insofar as institutionally disciplined documentary practices provide objectifying resources they also contribute to such projects. But legitimation of one form also works to delegitimate others. Documentation can reveal exclusions and delegitimations. The permeability of these borders is also apparent in nonexpert appropriations of scientific discourses.[2]

4 How documents figure in scientific practices varies across a spectrum from authorized to unauthorized science. At one extreme lie the stabilization projects of modern, authorized science. I have argued that science's formal documentation plays an essential role in such projects, by enacting a literary performance of the routinizations and standardizations of epistemic alignments by virtue of which local laboratory phenomena can become informative elsewhere. Yet the specific form that epistemic alignments take are not hard-wired, as it were, to translaboratory intelligibility. It is not the case that the only way a local phenomenon can be intelligible elsewhere is to submit to the routinizations and standardizations characteristic of the universalist and objectifying projects of modern, Western science. There are other processes of knowledge production, whose procedures of legitimation and sources of stability either ignore or exist in opposition to those of authorized science. Thus further along the spectrum we find appropriations by activist groups (environmentalists, AIDS activists, animal liberation movements, advocates of alternative medical treatments, and others) of statements found in science's formal literature to challenge conclusions issuing from what they regard as destructive institutional sites. The problem they face is not that their documents fail to represent nature, but that the legitimacy of the formal literature is not sufficient to overcome resistances of other powerful elements of authorized science. Still further along the spectrum, the significance of a literary technology of objectivity withers away altogether in knowledge traditions reliant upon craft and apprenticeship relations for delocalization and trans-situational intelligibility. The success of their efforts to document knowledge of the natural world in unauthorized and illegitimate contexts depends not their succeeding or failing to convey 'pure information' about nature, but upon the strength and durability of the cultural resources available to them. A

research program in scientific documentation opens investigation to alternative traditions and procedures through which knowledge of the natural world may be gained, and of the different roles played by the production and use of documents within them.

Document Strata

Fleck drew attention to the several documentary strata of scientific labour. Because he regarded the processes of objectification as embodied in the material practices of writing, he spoke about the central role of special expressions, technical terms, and symbols; of whole sign languages (those 'used in chemistry, mathematics, or symbolic logic') and the importance of numbers; of the ways in which meanings and concepts are fixed in what he calls a 'lifeless language,' which renders them 'static and absolute' (1979, 144). Latour's work shows that laboratory labour is thoroughly immersed in masses of documents. In describing his need for documents to pursue historical work on 'how arguments emerge from the modern physical laboratory,' Galison writes:

> Some of this need is fulfilled by notebooks letters, or published papers ... As I began to piece the story together, I found project proposals, progress reports, conference proceedings, computer outputs, data tapes, film records, flow charts, circuit diagrams, blueprints, scanning records, administrative correspondence, report transparencies ... Team members ... argue through meetings, and they summarise these in photocopied minutes; they make proposals and counterproposals through technical memoranda circulated only within the group; they put forward claims and refutations as the data begin to bubble up through the many-layered filters of analysis. Papers like these compose a new form of scientific literature that physicists have created in response to the explosive growth of collaborative experiments. (1987, x)

Schaffer also draws attention to the variety of documents in science and the varying practices with them: 'Working scientists make, handle, and translate representations through networks of machines and humans. Because of this process, such representations can take many, varying forms. Just as literary technologies are sociohistorically local, they are also specific to particular representational practices. Manuscript notes, copper engravings, daguerrotypes, chart recorder graphs,.

and computer printouts all take part in different regimes of representation and different labour processes' (1998, 183).

Many research problems emerge from the masses of documents involved in scientific labour. We get an idea of what is at stake in issues concerning the constitutive effects of document strata by the attempts of Fleck, Shapin and Schaffer, Ravetz, and Rouse to locate the documentary site of objectifying resources. Fleck locates certainty, vividness, 'the solidarity of all human knowledge' and the belief in the 'possibility of a universal science' in science's exoteric, or popular literature, and claims that a scientific statement first achieves sufficient stability to operate as a 'thought constraint' in the vademecum rather than the journal literature. Shapin and Schaffer, on the other hand, interpret the journal literature as an objectifying resource, whereas Ravetz finds the research report the site of robust objectivity. Rouse insists that no form of scientific literature includes 'an encyclopedic representation of the present state of knowledge in the sense in which it is typically reified by most epistemologies' (1996, 174), locating the objectifying process in the narrative function of scientific activities.

These sorts of debates are important, and they become more complex when the range of document types and their interactions are expanded. One set of problems consists simply in identifying the various strata of documentation and the conditions of their existence. What determines a document type – discipline, problem, organizational structure, or some other factors? What are the effects of new forms of documents on theory, experimentation, observation, writing practices, social relations, and the possibilities of building stable networks (for example, the new imaging and simulation techniques made possible by developments in computing)?[3] The growth of interest in the relationships between various kinds of documents and scientific practices is shown by recent work that includes, for example, historical studies of natural wonders and marvels (Daston and Park 1998), new perspectives on the role of print and book-making in the historical development of modern science (Johns 1998), the role of writing in mathematics (Rotman 2000), exploring the borders of science and art (Jones and Galison 1998), the role of graphic representations in science (Lynch and Woolgar 1990), the intertwinings of the history of science and the history of the book (Frasca-Spada and Jardine 2000), and case studies in the materiality of communication (Lenoir 1998).

History

The importance of history to studies of scientific documentation has emerged at various points in previous chapters. The reason for it follows directly from the conclusion that documentary effects are historical phenomena. 'Literary technologies are sociohistorically local,' as Schaffer puts it (1998, 182). A proper analysis of them shows how they arise, are sustained, and pass away. Studies of scientific practices show that the 'de-localization' of scientific phenonomena that manufactures objectivity and universality is the product of historically contingent forms of scientific labour. Such labour includes equally contingent forms of scientific documentation. Historical investigations of the processes by which scientific phenomena are stabilized become, to borrow the title of a book by Shapin (1994), social histories of truth. When expressed in the idiom of information, analyses of the contribution of scientific documentation to such stabilization become social histories, not of the *communication* of scientific information, but of its *constitution*. The examples in chapter 6 illustrate this point. They help show the relevance of historical investigations to explanations of the role of inscriptions and documentary practices in the construction of scientific facts as objective and universal.

From the strange particulars of Baconian science to Newton's prescient attempts to ground facticity in highly theorized representations of the natural world, historical studies of the role of inscriptions and documentation in constituting scientific facts show that whatever 'information' might be thought to be 'communicated' by and through such inscriptions and documents is an effect rather than the cause of the stability of scientific claims. Stability is not a product of representations that fill in the gaps of an incomplete picture of nature, but of the complex and historically variable interactions between various literary, material, and social technologies. Historical studies also reveal the many differences in the ways scientific practices are open to those inhabiting other regions of the cultural terrain. The historical work of Dear, Shapin and Schaffer, Daston, and Bazerman show that the documentary practices supporting the stabilization of scientific facts shift in different historical and cultural periods. These are not shifts in the type or quantity of scientific information, nor in the velocity of its communication, but in the conditions of the possibility of the informativeness or communicability of specific kinds of documents.

The Fantasia of the Science Library

So far, the contribution of scientific documentation to the manufacture of knowledge has been considered from the perspective of science's formal literature. But documents are not available to scientific labour in an unorganized and undifferentiated heap. Instead, they are subjected to a highly institutionalized social discipline of subject organization manifested in the numerous technologies of information retrieval – for example, various kinds of controlled- and uncontrolled-vocabulary subject access systems, thesauri and classification schemes, search engines with syntactic and semantic capabilities of varying degrees of sophistication, systems of document and query matching based upon linguistic principles or word frequency distributions, and clustering techniques based upon document properties such as bibliographical references, citations, or words in titles, abstracts, or full text. Moreover, the subject organization of scientific documents has an objective correlative in the ordered arrangement of books and periodicals on the shelves of science libraries, which offers an impressive material metaphor of a massive and robustly stable organization of scientific knowledge. This section argues that the strength of scientific documentation's ability to function as an objectifying resource is greatly augmented by the scope, depth, and sheer inertia of these organizational apparatuses, their institutional settings, and their disciplined routines. Science's organizational documentation – the penumbra of subject heading and indexing systems, thesauri, classification schemes, and increasingly complex retrieval systems for text, music, still and moving images, and digital documents that surrounds its primary literature – belongs to the 'mass of documents' that Foucault argues is central to the stability required for different cultural and historical configurations of 'regimes of truth' and objects of knowledge. Just as, according to Foucault, eighteenth- and nineteenth-century documentary practices were central to the construction of 'man,' whereby knowledge of persons in their individuality became possible, so too do the documentary practices of science's organizational literature contribute their inertia and mass to the stabilization of scientific phenomena and to the knowledge of science's conceptual organization.

Fleck speaks of the 'mood of the thought collective of natural science,' which is 'realised in a particular inclination to objectivize the thought structures it has created.' He remarks on the 'tendency to reify and objectivize the conceptual creations of scientific thought,' and how it is 'inseparably bound up with' the 'migration of ideas throughout the collective' (1979, 144). The objectifying mood is located, he argues, in the exoteric literature, which provides the scientist with a frame of truth in the form of a representation of the natural world. Science's organizational documentation belongs to this exoteric literature. It contributes to this mood and this tendency, which Fleck describes as effects of depersonalization of scientific labour and its detachment from locality, particularity, and contingency. The organizational literature offers material support for the image of discrete units of epistemic content of documents fixed in comprehensive conceptual networks. The following sections develop the theme of the importance of science's organizational documentation to its objectifying project.

Keeping Records

Traweek's remarks about scientific writing as 'a record-keeping device' introduce a key concept for analysing the discursive labour expended upon the construction and maintenance of science's organizational apparatus:

> written materials, articles and preprints, are the commodities the physicists produce ... Articles represent the consensus, the 'facts,' data with the noise removed. The authors of these written accounts own the information in the account. Any subsequent users of that new information must pay royalties to the authors in the form of homage or credit, thereby increasing the accumulating reputations of the authors. In talk physicists rarely give credit to others. Scientific writing keeps track of the results of these debates. It is a record-keeping device, a spare ledger of credits and debits. Citations are a trace of something happening elsewhere, as a bubble chamber photograph presumably records traces of events in subatomic nature. (1988, 122)

Traweek's economic metaphor is a staple of analyses of scientific writing. It is implied in Merton's emphasis on reward for adherence to the norms of science; it is found in Warren O. Hagstrom's (1965) notion of writing as gift-giving, in Pierre Bourdieu's (1988; 1991) concept of accumulation and expenditure of symbolic capital, in Garvey's (1979) conception of articles as priority claims to intellectual capital, in Latour and Woolgar's (1986) cycle of credibility, and in managerial models of merit for scholarly productivity, such as the systems of academic accountability so enthusiastically embraced by Cronin (Cronin and Overfelt 1994; see also Cronin 1984). The metaphor keeps pace with our globalized network economy in James R. Beniger's (1988) interpretation of citations as a solution to a crisis of control.[4]

Traweek sees citation networks as elements of an accounting system. Any system of accounting presupposes the possibility of counting, and for things to be countable, they must be the same in some respect. Counting practices therefore depend upon established, authorized, and legitimated disciplinary practices that constitute criteria of identity for countable elements. Thus for the implicit network of citations to function as an accounting system there must be an accepted criterion of identity for what constitutes a unit of credit or debit. Each journal article, and each reference to another journal article, must count as a unit. Traweek's record-keeping device or accounting system thus rests upon criteria of identity that define scientific documents as individual units. A document can emerge as an individual or single unit – one like all the others in respect of its countability – only because it is embedded in practices that abstract from the respects in which it differs from other scientific writing. Bibliographic practices do the required work. They are highly institutionalized, disciplined, and socially regimented practices that make possible the knowledge of documents in their individuality, discrete and countable. Their documentation belongs to science's organizational apparatuses, technologies, and social disciplines, which constitute the organization of scientific writings in subject analysis systems and in libraries. With such practices firmly in place there emerge the conditions of the possibility of accounting systems such as those indicated by Traweek. Articles, authors, institutions, and journals can be ranked, sorted, and compared. Measures of central tendency and analyses of variance can be computed. These statistical regularities can be interpreted in different ways – among them as 'spare ledgers of credits and debits' – to support a wide range of conclusions, recommendations, policies, and actions.

The analysis of citations as 'a trace of something happening elsewhere' suggests a role for this 'record-keeping device' analogous to Knorr-Cetina's concept of the role of the journal article as a 'script of resource conversion.' Insofar as citations to articles and their bibliographical references generate statistical regularities, they can be interpreted as revealing the inherent properties or the structure of a literature. Statistics is a powerful legitimating device for translating abstract statements about relationships between numbers into statements about a world populated by countable things. Bibliometric studies are *scripts of interiorization*, because they provide resources for generating and stabilizing statements that describe the knowable interior of literatures. There may be other methods of coming to know that interior, there may be statistical regularities other than those provided by bibliometric studies, there may be criteria of identity for countable units other than those provided by bibliographical practices, and there may be disputes about the internal structures represented, whether by bibliometric studies or other techniques. But the objectifying trope of theoretical knowledge of a literature's properties and structure is underwritten by robust and stable bibliometric regularities. Bibliometric studies are scripts of resource conversion because they project the documentation of bibliometric, statistical regularities onto the documentation of theoretical knowledge of a literature's structure. They are scripts of interiorization because they construct a difference between a trace and data that generate it. Like any mathematical technique, they do not determine their own interpretations. It is not obvious, for example, whether the co-citation clusters produced by Eugene Garfield, Michael M. Kessler, and Henry Small represent, as they claim, the conceptual areas, or 'subjects' of scientific research, or whether they reflect the social networks of research work. Although the paths to knowing a literature may be multiple, bibliometric studies provide a script for stabilizing statements with a claim to represent a literature's internal properties, processes, and structures – 'as a bubble chamber photograph presumably records traces of events in subatomic nature.'

The idea of scientific writing as an accounting system ranges far beyond applications of the statistical techniques of bibliometrics. Even Latour's analysis of the scientific journal literature, although highly inflected by post-modern motifs, assumes a stable accounting system for articles. The descriptions of writing in *Science in Action* depict the production of writing as a process of stabilizing statements such that they can be used as 'black boxes' for use in writing further articles, not as a process of communicating information. For Latour, the statements of the scientific journal article are more usefully conceived by analogy to building materials than as units of information. But any labour process of constructing highly specific and, more importantly, reliable material products from assemblages of ready-made parts depends upon clear criteria of identity by which those parts may be identified, their value determined, and their availability readily secured. The reliability of auto assembly, for example, depends upon a stable accounting system for auto parts. Reliability of scientific fact construction likewise depends upon a stable accounting system for the elements used in fact construction. Accounting systems for building materials underwrite guarantees of the reliability of the final products of both kinds of construction projects. Accountability, in the form of a systematic inventory of system parts, validates such guarantees because it permits the process of production to be reconstructed through access to the organized records of system parts in cases of product breakdown and failure. Thus a condition of the possibility of an orderly and trustworthy labour process of scientific fact construction, which involves the use of past literature to write current literature, is a stable, public inventory system for scientific documentation. The reliability of a finished product can be contested or supported through the use of this system to reconstruct its process of production. Articles positioned outside this accounting system are, to borrow Latour's metaphor once again, 'like a child without an escort walking at night in a big city it does not know: isolated, lost' (1987, 33).

Organizing Documents

While the structure of citations and references constitute one of the accounting systems of scientific articles, the document retrieval apparatus of subject representation and organization is even more public, more deeply institutionalized, and involves even greater expenditures of time, effort, and both human and capital resources. The keywords, descriptors, thesaurus terms, classification headings and notations, and specialized indexes of subject organization systems belong as much to the documentary apparatuses of science's objectifying resources as do sentences, charts, tables, graphs, diagrams, photographs, and new modes of digital representation. The argument that the literary technology of science must be included in the processes of stabilization of scientific facts and phenomena does not restrict the relevant discursive elements generated by the labour of scientific writing to statements in propositional form. Statements should be understood in the Foucauldian sense, which includes non-propositional forms. Foucault's example of a statement makes this point most emphatically: the 'series of letters, A, Z, E, R, T, listed in a typewriting manual, is the statement of the alphabetical order adopted by a French typewriter'

(1972, 86). The processes aimed at stabilizing scientific phenomena through the characteristic practices of writing science's formal literature are therefore not limited to operations on propositions, as assumed in *Science in Action*, which interprets stability in terms of changes in propositional modalities. Keywords, descriptors, index terms, and the other discursive objects of the organizational apparatus of scientific documentation are not propositions, but they nonetheless belong to the set of *statements* made available by science's literary technology to stabilize facts and help construct the objectivity and universality of scientific phenomena.

The complex subject organizations of scientific literature are similar to the statistical regularities of bibliometric studies in at least one respect: stable regularities of system elements also support alternative interpretations. An element of a subject analysis system, such as an index term, is applied according to a specific criterion of identity, usually expressed as the subject of a document or of some unit of text within it. The number of documents marked by the same system element can then be counted, comparisons can be drawn, distributions of system elements can be modelled mathematically, and variations between systems can be quantified. For subject analysis systems as for bibliometric analyses, there are different answers to the question, What do these regularities mean? An opposition familiar to classificationists is pertinent in the case of scientific documents, since on one view the regularities of index terms are artifacts of their literary warrant, thus reflecting agreements and uniformities in the set of organized documents, while on an alternate view they are seen as reflections of the concepts and conceptual relations that pre-exist and are instantianted in documents. Yet on either view, the regularities of system elements issue an interpretive licence to speculate about what these regularities might *represent*, whether Society, for those championing literary warrant, or Reason, for champions of realist accounts of concepts and propositions. In either case, the stabilities of subject organization systems that enable document retrieval function as objectifying resources that support the objectivity of underlying regularities, whether interpreted as those of social or conceptual worlds.

The importance of these systems does not depend upon specific thresholds of use. The accounting system of scientific articles is as necessary to the labour processes of fact construction as inventory systems are necessary to the retail trade. Just as customers can find their way about department stores by following their own paths, scientists and other document users are known to find their way from one document to another without making significant use of system elements. LIS use studies confirm an inverse relationship between levels of scientific expertise and uses of subject headings, descriptors, classification and indexing systems, catalogues, libraries, and librarians. For experts, the discursive resources purporting to denote fixed concepts are known to be unstable, a point made by Rheinberger in his analysis of laboratory labour as the production of unexpected events.[5] Yet with scientific labour, as with the retail trade, inventory control or accounting systems ensure the orderly and coherent availability of stock by already organizing the terrain upon which the possibility of individual navigations through the documentary universe depends. The alternative to system is a crisis of confidence, because in the absence of an organizational apparatus, trust in the orderly availability of reliable raw material is undermined. The accounting system can be ignored only if it is already taken for granted because it continues to work silently and unobtrusively in the background. Just as daily financial transactions depend upon the regulation of currencies and commodities, so too does the circulation of intellectual capital depend upon the regulation of documents by systems of subject representation, organization, and classification. Once the system is in place, it can recede from the consciousness of specific actors whose transactions, conducted according to their own purposes and devices, nonetheless presuppose the *organizability* of documents, of which existing subject organizations are the most telling material metaphor. Thus the structure of scientific documents, and all the effort of representing, organizing, classifying, collecting, and processing – in short, all the processes of building a *library* – are central to the availability and the meaning of the scientific journal article, whether considered as raw material for the construction of facts, or as a commodity in a marketplace of intellectual capital, whether or not individual scientists make little explicit use of the organizational apparatus in their 'information seeking behaviours' or their 'information needs and uses.'

Once again, Fleck's analysis of the epistemological function of the exoteric literature of the natural sciences is a useful resource. It provides a model for thinking about the manner in which the objectifying function of resources such as retrieval systems can be central to advanced scientific work even though measures of their use by scientists are low. Fleck shows that although the literary activities of scientists centre upon debates in a highly unstable journal literature, nonetheless the stabilities of the exoteric literature provide an objectifying framework for their writing and reading. Any particular fact may be contested, but the cultural form of legitimate and institutionalized contemporary science demands that the discursive products of formal writing aim at the stability many come to enjoy in the exoteric literature. The exoteric literature engenders the objectifying 'mood' of the scientist. Similarly, the stabilities of retrieval systems and their material instantiations in the library depend upon a project of writing in which the product – the journal article – adheres to a discursive form sufficiently robust to permit itemization, classification, and subject analysis. Its structure allows a sufficiently clear criterion of identity to be applied in counting and classification techniques. Although the retrieval systems of scientific documentation may be 'mangled' by users embarked on their own journeys, the journal article is written in such a manner that its documentary form already projects a possible place in an organizational apparatus. This aim is analogous to the one concerning Fleck, that of writing statements such that they exhibit the form of facticity found in the exoteric literature. A condition of the possibility of even contesting the 'correct' position of a text in a subject organization of a scientific field depends upon its being constructed as organizable in the first place. The form of discreteness that positions an article in a subject organization is the same as that which permits the article's statements to lay claim to representing stable phenomena of an objective world. The universalist, representationalist, and objectifying imperatives of modern and contemporary science not only give meaning to laboratory results as stable facts of an objective world, but also make possible the emergence of epistemic content of textual units within articles, which can then stake claims for fixed positions in highly inert and stable systems of subject representation.

Once absorbed or translated into the subject analysis components of document retrieval systems, the terms used in journal articles undergo a decontextualization beyond the journal article's typical de-localization of laboratory contingencies. As they become subject retrieval elements of large document sets, they become abstracted from their original contexts. The stability of this 'document content' increases as a function of the sheer size, organizational complexity, and inertia of the subject organization systems in which it is now deeply embedded. Thus the systems of scientific subject organization provide, through their uniform, standardized representational elements and their more-or-less complex classificatory relationships, an important set of objectifying discursive resources through which science itself is thought and imagined as a stable and coherent conceptual field. Such systems take their place among the 'mass of documents' that fix and stabilize inscriptions to the degree that they can become available to scholarly investigation as 'scientific information.'

Building Libraries

Subject organization systems for document retrieval are central to the comprehensive science library. Science libraries house the exoteric literature identified by Fleck as central to the expert's faith in the objectivity of science, the esoteric journal literature, and its supporting organizational documentation. The conceptual networks of classification systems are embodied, in the row upon row of organized volumes, in the cascading lists, in volumes of finding aids, in interrelated subject headings, index and thesaurus terms, and classification notations, and in the organizational structures and work routines that configure professional practices with these document organization technologies. And far from deconstructing stabilities of typographical fixity, the transposition of these systems into electronic databases, the Internet, and the World Wide Web simply reify their discursive elements in an abstract conceptual space by virtue of the digital medium's embrace of mentalistic tropes.

If the journal article belongs to a first-order objectification of scientific rationality and order, and science's organizational apparatus to a second order, the science library may be counted as a third-order objectification. The library objectifies by fixing places for documents on library shelves, and through the operation and maintenance of a subject analysis system that maps decontextualized epistemic content to arrays of documents in a highly privileged and culturally respected institutional space. But the route to the library's institutional materialization of conceptual order begins with the journal article. The article's already decontextualized narratives supply abstract, decontextualized inscriptions readily translated into the stable terms and relations of science's subject organization systems. Insofar as the library multiplies material representations of unified and coherent organizations of scientific documentation, it becomes a tool for thinking and imagining science itself as a field of stable, abstract, decontextualized 'information' about a correspondingly stable and objective natural world. The science library's institutional instantiation of scientific information is a material support for an objectifying discourse of natural science as a stable conceptual field.[6] As the institutional site for the management of scientific documentation, the science library joins the journal article and its subject organization apparatus as a powerful objectifying resource of scientific documentation.

Foucault's 'Fantasia of the Library' helps us understand how the science library helps us imagine science as an organized conceptual field. Foucault argues that Gustave Flaubert's *Temptation of St. Anthony* is 'the first literary work whose exclusive domain is that of books' (1977, 92). Although the phantasmagoric *Temptation* appears to be 'the product of dreams and rapture' (89), or a 'property of the heart' (90), Foucault shows that it is instead 'a monument to meticulous erudition' (89), the product of 'the scholar's patience,' 'untiring attention, zealous erudition, and constant vigilance' (90), deriving from 'words spoken in the past, exact recensions, the amassing of minute facts, monuments reduced to infinitesimal fragments, and the reproductions of reproductions' (91). The *Temptation* is not what it presents itself as being – the literary expression of a frenzied imagination or religious hallucination. Instead, it is 'a phenomenon of the library' (91). In writing the 'first literary work to comprehend the greenish institutions where books are accumulated and where the slow and incontrovertible vegetation of learning quietly proliferates' (92), Flaubert 'was responding to an experience of the fantastic which was singularly modern and relatively unknown before his time, to the discovery of a new imaginative space in the nineteenth century' (90). In this new imaginative space, 'fantasies are carefully deployed in the hushed library, with its columns of books, with its titles aligned on shelves to form a tight enclosure, but within confines that also liberate impossible worlds' (90).

Foucault's discussion shows that, beginning in the modern period the library becomes the site for the production of what, despite appearances, turns out to be a rigorously constructed 'fantasia.' But I think it would be a mistake to restrict his point only to archival erudition in the production of phantasmagoria and fantasy. I believe the point is deeper, because it can also illuminate the production of a *fantasia of science*. This more fundamental notion of the fantasia of the library emerges from Foucault's comparison of Flaubert to Eduoard Manet. He says that both 'erect their art within the archive' (92). Manet's painting recognized 'the new and substantial relationship of painting to itself, as a manifestation of the existence of museums and the particular reality and interdependence that paintings acquire in museums' (92). Analogously, Flaubert's writing recognized new relationships among texts as manifestations of the existence of libraries. But these new relationships generate fantasia – imaginary products that disguise what they owe to 'the greenish institutions where books are accumulated.'

Scientific writing is not so different. The 'particular reality' of the scientific journal article is not the local context of laboratory labour, but a decontextualized information space of fact, observation, and theory. It is a space in which statements vie against each other in contests of stability. The rewards of stabilization consist in a statement's transcendence: it frees itself from its scene of origin and migrates to others, where it is deployed in further contests of stability. These contests, and the narrative structure of scientific practices, connect statements to each other. The science library embodies these networks of connected statements, in its organized collections of inscriptions and systems of subject representation. Thus scientific writing is a manifestation, no less than Flaubert's *Temptation* and the texts it made possible, of the existence of libraries and the 'particular reality' and interdependence its texts acquire in them.

Foucault argues that Flaubert's phantasmagoria emerge from the nineteenth century's new relationship of texts to each other, rather than from any representational relationship between text and the unique mental states of the creative, even frenzied, imagination. I have argued that the objectivity of the natural sciences emerges at least in large part as the effect of stabilities brought about by institutionalized practices of scientific writing. Such practices generate objectivity from relationships between statement and practice, statement and statement, and text and text, rather than from any representational relationship between them and objective phenomena of the natural world. Insofar as the relationships between texts are manifested by the existence of libraries, the objectivity of natural science is as much a fantasia of the library as the phantasmagoria of the *Temptation*. Foucault's point about the 'impossible worlds' liberated by the confines of the 'tight enclosures' that organize books on library shelves holds equally for the worlds of the scientific imagination: 'fantasies are carefully deployed in the hushed library, with its columns of books, with its titles aligned on shelves to form a tight enclosure, but within confines that also liberate impossible worlds' (90).

Gary Radford's reading of Foucault's fantasia of the library interprets 'impossible worlds' as 'worlds other than the objective world constituted in the discursive arrangements of science' (1992, 420). But

Foucault's use of the phantasmagoric and fantastical writing of the *Temptation* to support his claim about the nineteenth-century turn in modes of creative literary production does not exclude applications of his insight about the relationship between fantasia and the library to other kinds of writing. Since, for Foucault, all knowledge and truth emerges from both the dominance of and resistance to disciplinary discursive regimes, the space liberated by the 'intervals between books' becomes not the exclusive terrain of 'impossible worlds' as Radford interprets them, but are equally hospitable to scientific writing's construction of 'impossible worlds' from science's universal truths of nature. Scientific truth needs to be included in the kinds of truth indicated in one of Radford's own valuable insights: 'Truth is discovered not only in the library through the location of a particular text ... but it is also made possible by their arrangements and in the "spaces" that these arrangements make possible' (418). The imagination that produces nature operating according to universal laws is not as far from the phantasmagoric imagination as Radford supposes.[7]

Postscript: Science and Classical Music

The readers of this book are blessed by a culture that includes the phenomenon of classical music. The history of this phenomenon charts its extensive articulation throughout our social world. Classical music is a relay for circuits of performance, celebrity, scholarship, commerce, schools and universities, manufactures and industries, and a wide range of arts and crafts. Its articulation comprehends diverse cultural ideas: conceptions of the solitary genius, social status, notions of taste, discernment, sensitivity, intelligence, and refinement. It intersects attempts to police the borders of elite and popular culture, and defines what creative defiance of them means (consider the Kronos Quartet's performance of 'Purple Haze'). One thread of this history traces the role of writing and masses of documents. Classical music is immersed in documentary practices to a degree unmatched by other musical forms. The possibilities of classical music, its history, and what it means to be an expert, involve reading and writing in a manner and degree not found in jazz, folk songs, blues, or rock and roll. Not only reading and writing musical notation, but also the mass of scholarly writing, the works of critics, the archives of composers, and popular representations in a variety of media forms all play an important role in shaping this cultural phenomenon. There is a great deal of music without writing, but not classical music. Writing and masses of documents are central to this cultural phenomenon. But it need not have been so. Indeed, there need not have been any music at all. Yet if we are to accurately describe the cultural phenomenon of classical music, we cannot ignore how it has been and continues to be configured by writing.

The main argument of this book can be summarized by saying that the natural sciences are much like classical music. First, like classical music, they are articulated throughout our social world. Latour and the actor-network theorists, and the many writers on the social construction of technology, have traced these connections in detail. Second, like classical music, science is a contingent cultural phenomenon. There need not have been what we call science. And what science there has been, has been thoroughly immersed in the same kinds of social practices as those that configure classical music: the practices of writing masses of documents. Just as there is music beyond writing, there is knowledge of the natural world beyond writing. And, just as there is music other than classical music, there is much knowledge of the natural world without the disciplined kind of writing prescribed by the literary technologies of authorized, Northern/Western science. But the dominant form of scientific work institutionally legitimated as 'real' science depends, like 'real' classical music, on institutions of writing and masses of documents. The universalizing project characteristic of this form of knowledge is conducted to a large degree through the objectifying resources of scientific documentation. If we are to accurately describe the cultural phenomenon of the natural sciences, we have to describe the way they have been and continue to be configured by imperatives of documentary practices. Our histories will include the various forms such practices take and the transformations they have undergone.

The role of documentary practices in the sciences is obscured when our gaze is fixed on the communication of scientific information. Information is central to epistemological analyses of science which see scientific work primarily as the cognitive labour involved in building conceptual networks. Theory building plays a key role in these intellectualized analyses. But studies of scientific practices show how far such analyses drift from what actually happens in the laboratory, and in scientific labour generally. The closest that scientific practices come to the kind of intellectual labour articulated in epistemological discourses of science is the production of science's primary literature.

There, the contingencies and localities of scientific work are thoroughly erased, and the thickets of its opportunistic reasoning are paved over by the smooth highways of a scientific method that aims at theory production. Only then do paradigmatic forms of 'scientific information' begin to emerge. It should therefore come as no surprise that when information science investigates scholarly communication among scientists, or their 'information-seeking behaviour,' it finds that these documentary circuits are not the primary resources for research-front work. Insofar as science mandates pursuit of universalizing and objectifying projects, the disciplines of writing, the masses of documents, and their institutionalized apparatuses of organization and subject analysis become crucial to its historical and cultural configuration. Writing and documentary practices are central to it in ways that information and communication are not.

Notes

Introduction

1 The reviews of research on the uses of information appearing in the *Annual Review of Information Science and Technology* show that sustained attention to this area of study by the academic discipline of library and information science began in 1963 with studies of the uses of scientific information. These reviews are discussed in chapter 2.

2 The production of a published paper has often been used as a criterion to distinguish the work of scientists from that of engineers. Herbert Brinterg, quoted by Thomas E. Pinelli, notes that '[u]nlike scientists, the goal of the engineer is to produce or design a product, process, or system; not to publish and make original contributions to the literature' (Pinelli 1991, 12).

3 Information science is no exception; apologetics for the discipline's lack of robust theoretical knowledge have taken the form of pleas for tolerance of an 'immature' science, likened to physics and chemistry in their infancy (Hernon and Schwartz 1993).

4 The discussion throughout this book follows an accepted convention of treating the journal literature as the primary literature of science. Whereas preprints, notes, letters, conference presentations, laboratory notebooks, and so on represent the raw material from which the journal article is produced, the article itself is the first complete, vetted, and finished presentation of scientific results. The secondary literature, found in manuals, handbooks, or what Ludwik Fleck called 'vademeacum science' (Fleck 1979), represents the selection and codification of such results and a higher degree of consensus. Tertiary literature refers to the further condensation, refinement, and synthesis into encyclopedias, specialized glossaries and dictionaries, and similar reference works. Science's popular literature runs

the gamut from serious reflections by scientists upon their work for an educated reading public, to scientific journalism, which itself exhibits a spectrum from serious reportage to 'pop science.' Fleck insisted upon the
epistemological importance of popular science in the 1930s.

5 The work on 'citation maps' in the 1970s by Eugene Garfield, Morton
Malin, Henry Small, and Belver Griffiths mined the journal literature for
just such representations of the current state of scientific research. Citation
patterns, particularly patterns of co-citations, were thought to provide
time-sensitive maps of science's conceptual patterns. Thus Small wrote: 'If
it can be assumed that frequently cited papers represent the key concepts,
methods, or experiments in a field, then co-citation patterns can be used to
map out in great detail the relationships between these key ideas' (1973,
265–6). The theory was also championed by Garfield: 'the citation is a precise, unambiguous representation of a subject that requires no interpretation and is immune to changes in terminology' (1979, 3). Thus for some
time, 'research fronts' automatically produced by co-citation clusters functioned in the SCISEARCH database as index terms used in subject searches
(Garfield 1981a; 1981b; 1981–82). See also Garfield, Malin, and Small 1978;
Small and Griffith 1974; Griffith, Small, Stonehill, et al. 1974.

6 Elizabeth Eisenstein (1979) makes the best-known argument for the dependence of modern science on the reliability – the 'typographical fixity' – that
print and the printed periodical article provide, for robust and effective systems of communication and information transfer. For a sustained criticism
of her concept of 'print culture,' see Johns 1998.

7 The development of science's 'literary technology' is discussed in Shapin
and Schaffer 1985; see also chapter 6.

8 For a criticism of this view, see Latour and Woolgar 1986, 203–8.

9 Theoretical attention to information itself is not as popular in information
studies as it once was; for exemplars of the type, see Belkin and Robertson
1976; Brookes 1980a; 1980b; 1980c; 1981.

10 For a recent work that addresses the phenomenon of information from perspectives of critical theory and historical studies, see Day 2001.

11 G.P. Baker and P.M.S. Hacker provide a useful summary of the philosophical importance of the Augustinian picture (1983, 1–27).

12 Karin Knorr-Cetina (Knorr Cetina 1999) refers to epistemic cultures, illus-
trating the concept through an investigation of the differences in the cultures of knowledge production in high-energy physics and microbiology.
The production, uses, and effects of documentary practices are emphasized
by Foucault (1979), where he notes the masses of documents produced to
support the surveillance of many different populations – prisoners,

patients, students, soldiers – from which truth about *individuals*, or persons knowable in their individuality, emerges as an effect. Foucault's familiar point about micropractices as the relays of institutional power applies to the documentary circuits that fix individuals as knowable, for incarceration, for medical treatment, for education, and for military discipline. This connection is discussed in chapter 4. (Knorr-Cetina has published under different forms of her name. In this book, she is referred to as 'Knorr-Cetina' in the text except where necessary to refer to alternate forms of her name in Works Cited.)

13 The ellipsis recognizes that Nunberg's question is more specific; he refers to 'practices of reading and the particular representations that support them.' Here, we are interested in the practices with scientific documents.

14 The idea of epistemic content assumes that 'the content of knowledge (both what is known and the evidence and reasoning that warrant it as known) [is] independent of its particular realizations in texts, utterances, or thoughts and of the specific history through which it came to be known' (Rouse 1996, 180–1).

1. Epistemic Narratives of the Social Life of Science

1 The details of Bacon's classification must be pieced together from several writings, since the Division of the Sciences was never completed. Bacon offered in its place a Latin, reworked version of his much earlier *Advancement of Learning* (1605). For an account of the texts and the details of Bacon's classification, see chapters 13 and 14 of Anderson 1948.

2 The importance of documents, their collection, organization, and the social structure required for the production of knowledge from them, is emphasized in Julian Martin's (1992) study of the relationships between Bacon's view of science and his design for the British imperial state.

3 Attention to the documentary production of scientific results in Bacon's scheme foreshadows Latour's (1987) 'centres of calculation.'

4 For a very good introduction to the relationship between knowledge and library classifications, see Miksa 1998.

5 For work on Paul Otlet, see Buckland 1997; Day 1997; Otlet 1990; Rayward 1967; 1975; 1991; 1994b; 1997; Rieusset-Lemarié 1997; *Transnational Association*, nos. 1–2, 2003.

6 Zagorin notes the persistence of Baconian ideas: 'Bacon's reputation as a philosopher of science stood very high in the seventeenth and eighteenth centuries, and was still considerable in the Victorian era' (1998, 90).

7 For a challenge to interpretations of Otlet as a positivist, see Day's (1997)

ingenious 'close reading' of Otlet's later work. Day develops some fascinating strands of Otlet's thinking about the nature of the book that demand further attention, especially ideas of the book as an 'accumulator' of mental energy, and as an 'assemblage of a multiplicity of forces, bodies, and senses.' But problems about Otlet's positivism remain. Twentieth-century positivism is a robustly anti-realist and anti-representationalist philosophical movement; it rejects as nonsensically metaphysical the chief idea Day assumes defines it: the idea of science as a representation of nature. Positivism is radically sceptical of a metaphysics that would link science's theoretical entities to reality by representation. Thus if Day is right in saying that for Otlet, 'the study of the book belongs not to the metaphysics of representation' (312), and that Otlet's thought rejects 'the notion of the book as a positive representation of reality from an "objective" position outside of that reality' (314), then in these regards, at least on Day's reading, Otlet and positivism actually agree. The task of locating a 'guiding metaphysics' for a thinker as non-systematic as Otlet is especially difficult, since many contradictory and competing strands of thought may be found in his writings. Yet his positivist allegiances are undeniably explicit in the early work. For a brief and accessible account of positivism in the context of debates about realism and representationalism in science, see Hacking 1983.

8 Otlet's *documents* are anything with evidentiary value, 'and so included objects themselves as well as graphic and other representations of objects and ideas' (Rayward 1991, 137). The willingness to treat *things* as documents was a defining feature of documentation as a new profession and scholarly enterprise (Buckland 1997).

9 The Library of Congress's rules for cataloguing books, for example, limit the scope of a subject heading to no less than a third of the book's content, and usually limits the number of subject headings per book to three.

10 For two recent studies of the spiritualization of digital information, see Davis 1998 and Wertheim 1999.

11 For a short and accessible introduction to the 'scientific ethos debate,' see Toren 1983; a useful list of references may also be found in Bazerman 1983, 168.

12 William D. Garvey also emphasizes the reward inhering in mere publication, through the 'use of journal articles as the primary source to establish priority' (1979, 75); 'in almost every scientific discipline today, the socially accepted medium for establishing priority is the scientific journal article' (69).

13 A paper's citations are the papers that reference it later on; a paper's references are those listed in its reference list, or bibliography.

14 The typology of uses of scientific papers that Bernal presented to the 1948

Royal Society Scientific Information Conference also assumed that published papers deliver information useful at the research front. The first of Bernal's four types of use was 'working use at the bench or in the field' (Royal Society (Great Britain) 1948, 56). But in his questionnaire study of journal use, Bernal found that 'a very small number (7 per cent) have been used at the bench' (596). To save the assumption, he notes that the data do not mean that articles read elsewhere might not have been wanted at the bench, but perhaps only that it was not possible to use them there.

15 The 'information-recognition exchange model of scientific organization' developed by Warren O. Hagstrom, Merton's student, according to which the 'organization of science consists of an exchange of social recognition for information' also exhibits the same dependency of the integration of moral and cognitive value on the logical role of the information exchanged for social recognition. In Hagstrom's analysis (1965, 12–23), when manuscripts are given as 'gifts' in exchange for the social recognition granted through publication, information of epistemic value to scientific knowledge is thereby rewarded only if the gift *is* information, that is, only if it is logically related to the derivation of new knowledge. Hagstrom makes this explicit in his comments about another form of recognition operating through the formal channels of scientific communication, the practice of citing the publications of others: 'It is usually necessary, even obligatory, for them to recognize previous work, *for the validity of their own contributions depends logically on the earlier work*' (24; emphasis added).

16 In a recent exchange between Stevan Harnad and Steve Fuller on electronic journal publishing, Harnad argues that the 'esoteric' literature, that is, scholarly journal articles, has no market: 'esoteric serial publishers will learn that their real clients are esoteric authors (actually, their institutions and granting agencies) rather than readers' (1995b, 311); the 'captive audience' of the journals 'is not the readership of the journals, it is the institutional library that must have the entire journal in hand for the few, if any, who ever consult any particular article' (317). In his response to Harnad 1995b, Fuller emphasizes the non-cognitive function of the journal literature: 'The communication of results, the allocation of credit, and the creation of an archive all reflect the *publicity* function of journals' (1995b, 300). For the final word of the exchange, see Fuller 1995a.

17 For a concise study of this shift, see Buckland 1996.

18 Many contemporary champions of cyberspace are not nearly as sensible as Bacon in this regard. For a critique of the occlusion of materiality and its imperatives in the rhetoric of a fantasized new digital information order, see A. Ross, 1998.

19 Bernal sought the aid of biological metaphors to convey the urgency of the situation to his distinguished audience at the 1958 International Conference on Scientific Information: 'I tend to see [the transmission of scientific information] as a complicated irrigation system which is continually fed from many sources and in which the individual plants (the users) depend upon what reaches them at any given time. Ideally, each should receive just the right amount of water at the right time, but in practice, owing to the sluggishness and irregularity of the system, some never reaches the plants in time, and much of it evaporates or runs into the ground on the way. At other times the flow of water is so abundant that the plants are waterlogged and cannot absorb what they need' (1959, 79). 'To use a biological simile, the method of transmission of scientific information is on the most primitive level of wind-blown pollen. The more pollen – above a certain level – the more miss the stigmata waiting to catch them. We ought to advance at least to the more selective stage of insect-borne pollination where with far less pollen more gets to the right flowers' (87).

2. Scientists and Other Information Users

1 For a study of Bush, see Zachary 1997. There are many sources documenting the militarization of American society in the post-war period. For a discussion in terms of the development of information technologies and the construction of human subjectivity as information processors, see Edwards 1997. See also Galison and Hevly 1992, and Greenberg 1999; 2001.

2 For a detailed account of the memex and the publication history of *As We May Think*, see Nyce and Kahn 1991.

3 See also Day 2001.

4 The search for legitimation has been an abiding feature of library science, and library and information science, since their first appearances on the map of scholarly disciplines. For a relevant study of the tensions between the two, see Bennett 1988; for a study of information science which makes explicit the ties between information science and the military, see Debons 1974. For anxieties about the profession from a gender perspective, see Harris 1992.

5 The idea of information as a theoretical kind derives from Rouse's (1996) discussion of knowledge as a theoretical kind. His ideas are discussed in more detail in chapters 3, 4, and 5.

6 One can of course hear a message in a language one does not understand, and one may 'read' only in the sense that one derives spoken words from writing, as in the case of phonetic reading. M.T. Clanchy points out that

'reading was linked in the medieval mind with hearing rather than with seeing.' Thus a medieval 'reader' need only hear a document read aloud by another: 'Neither the hearer of a book nor the *dictator* of a letter needs to be a master of every detail of the scribal technique himself, just as modern managers are not required to type or programme computers.' Clanchy observes that '[r]eading aloud and dictating permit the non-literate to participate in the use of documents, whereas reading and writing silently exclude the illiterate' (1993, 270–1). For a philosophical use of the concept of 'reading without understanding,' see Wittgenstein's (1958) discussion of reading in *Philosophical Investigations*, §§156–71.

7 Very few of the *ARIST* reviews fail to mention the daunting number of information needs and uses studies. Although not limiting her statement to research on science information systems, Crawford noted in 1978 that 'over the past thirty years, some 1, 000 papers on user behaviour and the use of information systems have been published' (61).

8 Richard Apostle and Boris Raymond (1997) use Lewis Carroll's idea of 'port-manteau' words to criticize what they call the 'information paradigm' in LIS. For a review, see Frohmann 1998.

9 For an introduction to the clustering techniques used in marketing, see Weiss 1988. For a Foucauldian, critical analysis of the role of clustering in the database construction of consumer identities, see Poster 1996.

10 A different, chronological version of models correlating scientific activity to literature types may be found in Michael Keresztesi's patterns of evolution of scientific disciplines (1982, 13–19).

11 Professional faith in the necessity of exhaustive searches seems to be an irreducible feature of the LIS imagination. Thus almost three decades after Egan and Shera, Keresztesi writes of the scientific method: 'Before this process can begin, however, a thorough search must be made in the accumulated records, that is, in the literature of the discipline, to see if this exploration has already been done, to avoid wasteful duplication' (8–9). Yet years before, Merton had already pointed out that not only were multiple discoveries of the same scientific result common but also, 'when it comes to basic problems which are far from routine and, once solved, will have far-reaching implications for further inquiry, duplication, triplication or a higher multiplication of effort may be anything but wasteful' (1973c, 379).

12 Bernal's proposal for serious studies designed to 'see how scientific research, undertaken without any information at all, would go in comparison with that using the whole battery of modern information services' provides a refreshing contrast to the librarian's abiding faith in a professional conception of 'efficient' use of information retrieval machinery. He pleaded

for reducing the effort spent in 'routine studies of user behaviour' in favour of those 'which should test our conceptions of the whole function of information' (1959, 93). In a somewhat similar vein, Stephen K. Stoan observed that the librarian's lament for the scholar's presumed ignorance of efficient research skills is based on a failure to understand the elementary distinction between research and library skills. Stoan asks the obvious question: 'if scholars do not know how to use a library, how do they do research successfully?' (1984, 100).

13 Pamela J. McKenzie (2003) finds the origins of the idea of information encounters in the work of Patrick Wilson (1977). Its most explicit formulation is often attributed to Sanda Erdelez (1997). The principles of ELIS are expounded in Savolainen 1995. McKenzie's allegiances to the notion of 'information practices' are explained in McKenzie 2003.

14 Following Karen E. Pettigrew (1999), McKenzie calls such an information-rich site an 'information ground.'

15 The narrative form of research work is analysed by Rouse (1996) and discussed in chapter 5.

16 See T. Wilson 1984. For a critique of the cognitive viewpoint, see Frohmann 1992.

17 In his analysis of the 'phenomenology of information,' Nunberg writes, 'any effort to try to extract a coherent conceptual structure for the notion would be not just futile but false to its phenomenology: "information" is able to perform the work it does precisely because it fuzzes the boundaries between several genetically distinct categories of experience' (1996, 110).

18 There are many sciences in which experimental work does not figure at all. These are not the focus of the present study because the experimental sciences provide our cultural model of scientific rigour and the high road to knowledge of the natural world. If the arguments presented here hold for the hard cases, they provide greater support for conclusions about knowledge generally.

19 Bazerman is quick to point out that although LIS research acknowledges the importance of reading among scientists, its understanding of reading remains naive: 'Although reading consumes a substantial part of a research scientist's working life, science studies have not looked very far into exactly what happens when a scientist reads and how this reading is precisely related to scientific activity. Macroscopic surveys have documented the amount of time scientists in different specialties read, what kinds of documents they read, and from which sources they identify documents they might read. But in these studies, largely driven by information science interests in improving accessibility to information, the process of reading

itself has not been considered problematic' (1988a, 235). The assumption
that reading consists in bringing content to mind helps explain this lack of
attention to the problems posed by reading.

20 A recent scheme for disseminating the results of biomedical research on the
Internet echo Bernal's proposals sixty years earlier: 'The director [of the
National Institutes of Health], Dr. Harold E. Varmus, said his proposal for
an electronic publishing operation, called E-Biomed, would speed the
progress of science by accelerating the exchange of information among
researchers and by vastly increasing access to it' (Pear 1999).

21 Latour and Woolgar make a similar observation; they note 'a tendency evident in some sociology of science to stress the importance of informal communication in scientific activity. For example, it has been frequently noted
that the communication of scientific information occurs predominantly
through informal rather than formal channels ... This is particularly likely
where there exists a well-developed network of contacts as, for example, in
an invisible college ... Proponents of this argument have often played down
the role of formal communication channels in information transfer, choos-
ing instead to explain their continued existence in terms of an arena for the
establishment of priority and subsequent conferral of credit' (1986, 52).

22 Scientific journal articles are known to have some lay uses, although these
are rarely studied. A use of this kind was dramatized in the film *Lorenzo's
Oil*. I am grateful to my colleague, Dr Gloria Leckie, for the example.

22 The distinction is Hacking's; see his 1983.

3. Epistemology versus Practice

1 A recent work on the impact of electronic media on scientific communication expresses this hope: 'The conditions that gave birth to the journal and
the book, and that made each in its time the ideal medium for communicating, have changed. In the digital world, information conveyance devices
that provide identification, transmission, and storage functions are no
longer discrete physical entities. The "virtual library," that focuses on
access, transcends publications, walls, and institutions. ... the crucial element is users' needs – relevance of the information, economy of effort in
getting it, and facilitating conceptual connections ... Today, we have the
technological potential for designing systems that meet those needs'
(Crawford, Hurd, and Weller 1996, 6).

2 There are many manifestations of the belief that seeking and using information must be a systematic activity. A popular version leans on cognitive science. If we can discover the underlying principles of cognitive processing,

it is often thought, it will help us optimize information systems. For a critique of this idea as it appears in studies of the subject organization of documents, see Frohmann 1990.

3 My 'discourse of practice' is modelled upon Pickering's 'performative idiom' (1992, 144) and Rouse's conception of scientific practices (1996, 133–57).

4 Bazerman's gloss on Latour's *Pasteurization of France* (1988) provides a vivid example of the disparate elements of a specific assemblage, or scientific network: 'a successful scientist marshals crucial allies into an unbreakable network so that Pasteur's success depended on the cooperation of the microbes, cows, hygienists, colonial physicians, farmers, government policymakers, journalists, industrialists, scientific colleagues, and many other actants who were robustly drawn together through Pasteur's procedures' (1994, 118). Fleck (1979) was the first to draw attention to the historical, social, and contingent features of scientific work, in a process that he as early as 1935 called 'the genesis and development of a scientific fact.' His emphasis on the role of literature is especially striking, and is discussed in later chapters.

5 Pickering's *The Mangle of Practice* addresses this point explicitly, showing how the principles of his analysis of material agency in science 'can be put to work outside the realm of science,' thereby implicating issues of 'technology and the workplace ... industry, production, capital, and labour relations (1995, 31). By drawing the broad social context of science into his analysis, Pickering enacts the third stage of Harry Collins's Empirical Programme of Relativism (1992, 130). An earlier study along similar lines is Latour 1988. Politics and power as crucial elements of cultural studies of science are also addressed in Rouse 1987; 1996.

6 Bazerman's call for attention to the micropractices of scientific writing has also been echoed in the LIS literature, albeit from the narrower perspective of the citation process. See, for example, Cronin 1984, 52–3, 86–7.

7 Foucault's (1972) concept of a *statement* is useful here, to direct investigations of the role of discursive objects in scientific practices away from the epistemic content of inscriptions and categories of representation in general. The relevance of Foucault's work to the issues of this book are treated in more detail in chapter 4.

8 Rouse addresses most of these issues (1996, 24–30). On deflationary accounts of knowledge he says, 'A deflationary account of knowledge is directly opposed to epistemological views that take knowledge to constitute a theoretically coherent kind ... The deflationist recognises ... a wide range of examples of knowledge but denies that they collectively constitute

a coherent kind ... Deflationists take the ordinary uses of the term "know" and the associated practices of justification, reliance, and criticism, to be valuable and generally unproblematic, while denying that they can be explicated as instances of a well-formed kind.' On representation, Rouse notes that 'one of the striking consequences of this deflationary account of knowledge is its undoing of any *general* category of representations,' whether these are 'located in people's heads, sentences, skills and practices, or forms of life.' The inclusion of practices in this list is intended to reject interpretations of them as practical manifestations of a knower's grasp of their underlying structure. On this epistemological view of practice, understanding how to make things work is grounded in the presence to a knowing subject of a representation of a practical rationale. This view, Rouse explains, sees practices as 'analogues to the knowing subject' by 'equating practices with rules, norms, or skills that knowers or agents embody or enact.' Regarding epistemic content, Rouse argues forcefully against the concept of the 'supposed immateriality and ahistoricity of epistemic content and the presumed unity and coherence of beliefs and desires within and among individual scientific knowers.' On communities, he points out the close connections between talk of 'the representational content of scientific knowledge, the transparent communication of that content, and the scientific communities whose identity is formed in part by their allegiance to common presuppositions.'

9 This example shows the artificiality of construing as an 'information encounter' something learned from direct observation of the world.

4. Studies of Scientific Practices

1 Gilbert and Mulkay are quick to point out that discourse and context are not independent. 'It is important to emphasize that when we use the phrase "social context," we are not referring to phenomena which exist independently of participants' discourse. For social contexts are themselves *products* of discourse. It is through their recurrent patterns of language-use that participants construct such phenomena as the "formal research literature" and "informal" interaction' (1984, 39).

2 For a critique, see Frohmann 1992.

3 The concept of *information* presents challenges even to the empiricist repertoire. Some studies have shown that when scientists are polled about how specific forms of 'information' help them at different stages of their research, even those most accommodating to the concept, such as the production of ideas, concepts, and theories, the response is often sufficiently

incoherent to lead some investigators to abandon it altogether (Bitz, McAlpine, and Whitely 1975). If questions are posed in representational terms, answers reinforcing epistemological myths about scientific labour should come as no surprise. Studies of what scientists say about their work reveal, at best, sets of discursive regularities in different contexts. If discourse is treated as a topic rather than a resource, it cannot avoid the problem of accounting for the differences and contradictions between them. Gilbert and Mulkay's work shows the naïvety of accounts that uncritically take what subjects say as a window on the world.

4 Duhem's doctrine supports views about the under-determination of theories by observation by showing that there is no direct high road from data or observations to theory. The idea that such a high road exists enjoys a revival in LIS's appropriations Barney G. Glaser and Anselm L. Strauss's grounded theory, where theory is constructed through the manipulation of data taken to represent participants' beliefs free from any prior theoretical contamination. It is also assumed that analysts encode no theoretical assumptions in their data-processing techniques.

5 Rouse notes that Cartwright and Kuhn 'present us with a picture of theory as a disjoint collection of models whose range of application is not fully specified and whose effectiveness and accuracy vary considerably within that range. Furthermore, the use of theory is more a practical matter of making ceteris paribus adjustments to a concrete model to fit the needs of a specific case on a particular occasion than it is in providing a formal, deductive derivation from general principles' (1987, 85).

6 Commenting upon the significance of Hacking's work, Rouse notes that he has shown the great variety of activities that fall under the term 'theory': 'Hacking has already noted that ... ['theory'] ... denotes a motley of things, including speculations, physical analogies, ad hoc mathematics, and the variety of models that articulate the world theoretically and make it calculable; much of scientific work also proceeds relatively independently of this motley of theorisings' (1996, 114). Elsewhere he writes: 'Theory, [Hacking] tells us, is not just one thing with one function. Theories may be sweeping, suggestive speculations that only point us in a general direction, or they may be elegant formal mathematical representations. They may be physical models that give us a qualitative understanding of causal interactions or more or less ad hoc mathematical representations ... One does different things with theories of these various types. They do not form a seamless web of belief. Their coverage overlaps, and they may provide inconsistent versions of the same phenomenon. Some phenomena may fall in the gaps between the various kinds of theory we have in a domain and are conse-

quently not well treated by any. Theory gives us not one "world picture" but a diverse range of representations and manipulations' (1987, 85).

7 The television series *Junkyard Wars* is perhaps the best exemplar from popular culture of this way of thinking about scientific labour.

8 Fleck emphasizes the uniqueness of both the products of the laboratory and experiments. He writes: 'really valuable experiments ... are all of them uncertain, incomplete, and unique. And when experiments become certain, precise, and reproducible at any time, they no longer are necessary for research purposes proper but function only for demonstrations or ad hoc determinations' (1979, 85).

9 'I speak of experimental microworlds rather than simply of experiments because experimental practice is typically organized around the *family of objects, settings, and controls* that constitute the microworld' (Rouse 1996, 129; emphasis added).

10 Analogous differences in distributions of 'sameness' have been noted between 'documents' of oral and literate traditions. The oral tradition of reciting folk tales relies on criteria for the identity of a *story* that differ widely from literate criteria that fetishize inscriptions (see Lord 1960).

11 The term 'technoscience' is used here in the sense intended by Latour, to designate the networks of heterogeneous elements and associations whose strength, rather than propositional content, accounts for the construction and stability of scientific facts (1987, 174). See also Shapin 1988a, 538.

12 Hacking summarizes this belief: 'If the treatment works, then the world or the truth about the world makes it work, and that is what we have found out in the laboratory and then applied to the world.' He calls this 'the metaphysical mistake of thinking that truth or the world explains anything' (1992, 60).

13 For more details on Bell's invention and several other examples of the temporality of technoscientific production, see Latour 1987.

14 Some scientific work is marginalized only to be rehabilitated as a resource much later on, as the case of Gregor Mendel illustrates (Rouse 1996, 186–7).

15 A shift in the location of information from text to mind does not avoid the challenges to the concept of epistemic content that studies of scientific practices present. The objection to epistemic content as complete and wholly present does not depend upon its textual or mental instantiation. And if we admit that what is 'in the mind' of the reader must have a closer relationship to the text than the ideas that may arise from being struck on the head, then the 'content' in the mind is at least somehow dependent upon the 'content' of the text. Both are representations, and so is epistemic content. Even if the relationship between a text and its mental representation were

ignored entirely, objections to the concept of epistemic content would still apply to an exclusively representational *mental* field.

16 Pickering explains this dance of agency in more detail: 'Tuning in goal-oriented practice takes the form, I think, of a *dance of agency*. As active, intentional beings, scientists tentatively construct some new machine. They then adopt a passive role, monitoring the performance of the machine to see whatever capture of material agency it might effect. Symmetrically, this period of human passivity is the period in which material agency actively manifests itself. Does the machine perform as intended? Has an intended capture of agency been effected? Typically the answer is no, in which case the response is another reversal of roles: human agency is once more active in a revision of modelling vectors, followed by another bout of human passivity and material performance, and so on. The dance of agency, seen asymmetrically from the human end, thus takes the form of a *dialectic of resistance and accommodation*, where resistance denotes the failure to achieve an intended capture of agency in practice, and accommodation an active human strategy of response to resistance, which can include revisions to goals and intentions as well as to the material form of the machine in question and to the human frame of gestures and social relations that surround it' (1995, 1–2; emphasis in original).

17 'Around machines,' Pickering notes, 'we act like machines.' 'The field of practices is routinized and disciplined, *machinelike*,' because the proper performance of machines 'requires adherence to a standardized sequence of gestures and manipulations' – 'the gestures, skills, and whatever required to set machines in motion and to channel and exploit their power.' 'Just as the material contours and performativity of new machines have to be found out in the real time of practice, so too do the human skills, gestures, and practices that will envelop them ... all of these aspects of disciplined human agency come together with the machines that they set in motion and exploit ... Just as material agency is temporally emergent in practice, so, necessarily, is disciplined human agency ... they are intimately connected with one another, reciprocally and emergently defining and sustaining each other. Disciplined human agency and captured material agency are ... constitutively intertwined; they are *interactively stabilized*' (16–17; emphasis in original).

18 Pickering explains that it is not generally the case that all elements that could be mangled will be mangled in every case: 'not everything is transformed in each and every mangling. Some cultural elements and strata are just not 'relevanced' in particular passages of practice ... not even all of the cultural elements that are positively relevant to given passages of practice –

that enter into intended relations of linkage or alignment – are necessarily mangled ... Certain cultural elements and strata do, then, endure through particular passages of practice' (206).

19 A free quark is one found in isolation. As Pickering explains, Gell-Mann and George Zweig accommodated the conflict between particles with third-integral charges and classical electrostatics 'by suggesting that quarks were always to be found bound together to form hadrons in just the right combinations such that the third-integral charges of the constituent quarks added up to an integral multiple of e' (72).

20 See Wittgenstein 1978, 54; 1979, 50–1.

21 Pickering emphasizes this point in two places: 'Nothing substantive in scientific culture or anywhere else ... necessarily endures through and explains the process of cultural extension; everything in scientific culture is itself at stake in practice; there is nothing concrete to hang onto there' (1995, 111–12). 'I want to stress that on my analysis *nothing* substantive explains or controls the extension of scientific culture. Existing culture is the surface of emergence of its own extension, in a process of open-ended modelling having no destination given or knowable in advance. Everything within the multiple and heterogeneous culture of science is, in principle, at stake in practice' (146).

22 See Wittgenstein 1958, §83.

23 Latour's work has emphasized the 'network of writing' in which scientific labour is engaged (1987; Latour and Woolgar 1986). For studies of standardized forms of representation in science, see Lynch and Woolgar 1990.

24 In two paradigmatic papers of the actant network theory, Callon (1986) attributes agency to the scallops of St Brieuc Bay in France, and Latour (1992) insists upon the agency of doors.

25 Rouse observes that Traweek shows that the 'disciplines needed to establish and extend laboratory practices and achievements ... include the habitual practices and skills through which people make themselves into competent, reliable participants in a more or less shared world ... these practices are typically situated in specific social and cultural contexts; acquiring a physicist's understanding of the world is not fully or readily separable from other aspects of one's socialization as a physicist' (1996, 132).

5. Literary Technologies of Science

1 The magical qualities of scientific writing are nicely captured in a scene from *Perfume: The Story of a Murderer*, where the perfumer Baldini desper-

ately attempts through writing to grasp the intuitive genius of his apprentice, Grenouille: 'In due time he ferreted out the recipes for all the perfumes
Grenouille had thus far invented, and finally forbade him to create new
scents unless he, Baldini, was present with pen and paper to observe the
process with Argus eyes and to document it step by step. In his fastidious,
prickly hand, he copied his notes, soon consisting of dozens of formulas,
into two different little books – one he kept locked in his fireproof safe and
the other he always carried with him, even sleeping with it at night. That
reassured him. For now, should he wish, he could himself perform Grenouille's miracles, which had on first encounter so profoundly shaken him.
He believed that by collecting these written formulas, he could exorcise the
terrible creative chaos erupting from his apprentice. Also the fact that he no
longer merely stood there staring stupidly, but was able to participate in the
creative process by observing and recording it, had a soothing effect on Baldini and strengthened his self-confidence. After a while he even came to
believe that he made a not insignificant contribution to the success of these
sublime scents. And when he had once entered them in his little books he
no longer doubted that they were now his and his alone' (Süskind 1986,
108–9).

2 Latour's semiotic view of scientific practices is elegantly described in Hacking's characteristically laconic prose: 'Latour is a bracing reminder of that
glorious Parisian world of long ago, the late sixties, when inscriptions were
the reality and text was substance' (1992, 35n). Shapin's review (1988a) crit-
icizes Latour's world as a paper world.

3 Ravetz writes that a scientific task 'is not accomplished by the production
of an object which has a certain function, and an appropriate shape
imposed in certain sorts of matter. Instead, it is a statement of a particular
sort. The apparent form of the statement can vary widely: it can appear as a
report of an experiment or observation, a description and analysis of a complex situation, or as a statement of a fact or of a law, or as an hypothesis,
theory, or model' (1971, 117). Regarding problems, he says: 'We may now
put our definition formally, and say that a scientific problem is a statement
(always partial and subject to evolution) of new properties of the objects of
inquiry, to be established as a conclusion to an adequate argument, in
accordance with a plan (specified to an appropriate degree) for its achievement' (134).

4 The salience of the journal article in the actor-network approach associated
with Latour's work is emphasized by John Law: 'What can the laboratory
send out that will withstand the battering of the world? ... the most important is the text. It is the text most of all that the laboratory uses to rebuild

the world. It is the text that above all else forces others, willingly or not, to consider the importance of drug-carrying polymers in the treatment of cancer. It is the text that boxes in and regulates the points of contact between clinicians and researchers, patients and fund-raisers, polymers and livers, laboratories and diseases. In short it is first and foremost the text which imposes a structure on the world' (in Callon et al. 1986, 49).

5 See especially Latour 1993; 1996; 1999.

6 Collins (1992) argues that the very idea of successful replication is tautological, because the criterion of a successful replication is the derivation of the same result.

7 Bazerman's paper is a useful introduction to some of the pre-1980 literature on the sociology of science and its implications for scientific and technical writing. He has made the role and function of the scientific paper a distinct research topic; see 1988b.

8 Fleck had drawn attention to this feature of scientific writing almost fifty years earlier: 'The summarised report about a field of research always contains only a very small part of the worker's relevant experience, and not even the most important ... It is as if the words of a song were published without the tune' (1979, 96).

9 Knorr-Cetina cites one scientist's comments on the typical absence of recipe-like steps in scientific papers: 'There is a problem, of course, if one wants to replicate a result or repeat a method. As a rule, however, one does something else anyway. Hence, it is not so interesting to know exactly why and how certain things were done ... ' (1981, 129).

10 Knorr-Cetina finds that 'Results and Discussion effectively denies the interdependence of methods and results which ruled laboratory reasoning by relating the results not to their process of production, but to other results' (1981, 123). In his 1963 BBC broadcast, 'Is the Scientific Paper a Fraud?,' Peter Medawar put it this way: 'The section called "results" consists of a stream of factual information in which it is considered extremely bad form to discuss the significance of what results you are getting. You have to pretend that your mind is, so to speak, a virgin receptacle, an empty vessel, for information which floods into it from the external world for no reason which you yourself have revealed' (1990, 228–9).

11 Medawar remarks that 'in the discussion you adopt the ludicrous pretense of asking yourself if the information you collected actually means anything; of asking yourself if any general truths are going to emerge from the contemplation of all the evidence you brandished in the section called "results"' (1990, 229).

12 Knorr-Cetina gives examples of such changes of modality: 'Instead of say-

ing that something "is" the case, we find that "it has been suggested as possible"; instead of "should" we find "could" ... Expressions like "mainly" are watered down to "usually" or "commonly," and the "good" solubility of PPC becomes merely "enhanced"' (1981, 102).

13 Haraway's (1996) use of this expression is ironic, as it serves to criticize what she sees as the gender blindness of Shapin and Schaffer's work, a point she makes about the sociology of scientific knowledge generally. For more on Shapin and Schaffer's work, see chapter 6.

14 Fleck saw the same purpose in the 'modesty' of modern scientific writing, referring to 'a specific caution characteristic of journal contributions,' saying that it 'can be recognized in the typical turns of phrase such as, "I have *tried* to prove ...," "*It appears* possible that ..., " or negatively, "It was not possible to prove that ... "' 'Such jargon,' he noted, 'serves to shift the "holy of holies" of science, that is, any judgement about the existence or non-existence of a phenomenon, from the individual worker to the solely authorized collective' (1979, 118; emphasis in original).

15 The implications of Dear's contribution to the history of scientific universality are pursued in chapter 6.

16 See Bazerman 1988.

17 Knorr-Cetina elaborates: 'In the transition from laboratory work to the scientific paper, the reality of the laboratory changed. We have seen the situationally contingent, opportunistic logic of research replaced by a generalised context of present and possible worlds, and the interest negotiations of particular agents transformed into a projected fusion of interests of technology, industry, the environment and a human population needing protein. We have seen the reasoned selectivity of laboratory work overruled by formulaic recitations of the doings which emerged from this selectivity, and the measured results of these doings purged of all traces of interdependency with their constructive creation. We have seen the indeterminacy of the laboratory reduced to the careful expression of scientific doubt which the paper allows' (1981, 130–1).

18 Chapter 6 shows that at different times and in different places, objectivity manifests historical variety; it is a historically contingent phenomenon.

19 A study by the Manchester Business School and Centre for Business Research concluded that the 'flow and use of information was seen as an aspect of the research process which is primarily focussed on the production of information' (Bitz, McAlpine and Whitely, 1975, 1).

20 For an insightful study of the porous borders between institutionalized research science and the different ways in which science is articulated in a broader cultural arena, see A. Ross 1991b.

21 Rouse's view echoes Knorr-Cetina's concept of resource relationships. She has also noted that 'we need not assume shared morals, cognitions or interests in order to account for social cooperation.' To understand the social cooperation relevant to the coherence of science, therefore, 'we need not assume that scientists respond favourably to some research results because they share with the authors the criteria of relevance and standards of evaluation which imposes a certain "rational" selection. It is sufficient to assume a *temporary* fusion of interests through which scientists respond to results as a resource convertible into results of their own. Social action is interconnected not because of what is shared, but because of what is transmitted from one locus of action to another, transformed, and reintegrated, or because of a continued *process of conversion* which consists of the circulation *and* transformation of social objects' (1981, 131; emphasis in original).

22 The power and wonder of the installations at the Museum of Jurassic Technology in Los Angeles derive in large part from a deliberate blurring of boundaries between the scientific and technological, on the one hand, and the aesthetic, on the other. The cyborg discourses surrounding contemporary computer technologies provide other examples of such blurring. Thus even if the science and technology of Stelarc's art installations embody solutions to scientific or engineering problems, they remain beyond the pale of science and engineering unless renarratized in scientific contexts.

23 Rouse elaborates: 'Any particular scientific project thus acquires significance only against the understood background of what one does in this sort of research and what might actually be said in response to it. Scientific claims are formulated and defended with respect to the concerns and possible objections of an indefinitely extended group of scientists to whom they are addressed. There is little concern to argue for them *outside this specific social context*. Scientific claims are thus established within a rhetorical space rather than a logical space; scientific arguments settle for rational persuasion of peers instead of context-independent truth. Or rather, what it means for scientists to argue for the truth of their claims is to attempt rational persuasion of their peers' (1987, 120).

24 Latour's description of scientific activities as played out in an agonistic field, commentators have often noted, places great stress on struggle and conflict. Latour explains, however, the historical reasons for his discourse of warfare: 'The similarity between the proof race and the arms race is not a metaphor, it is literally the problem of *winning*. Today no army is able to win without scientists, and only very few scientists and engineers are able to win their arguments without the army. It is only now that the reader can understand why I have been using so many expressions that have military

connotations (trials of strength, controversy, struggle, winning and losing, strategy and tactics, balance of power, force, number, ally), expressions which, although constantly used by scientists, are rarely employed by philosophers to describe the peaceful world of pure science. I have used these terms because, by and large, technoscience is part of a war machine and should be studied as such' (1987, 172).

25 Pickering has also drawn attention to a related research area, in his comments about the need to understand more about the ways in which disciplined human performances in science are detached from particular scientific sites (1995, 109–10). In a footnote (110, note 39), he identifies an issue relevant to the general research area indicated here. Summarizing studies of scientific training and apprenticeship, he says: 'An interesting theme that emerges from such studies concerns the mangling of students' practices around pedagogical texts (manuals and so on) that are treated as at once authoritative fixed points and in need of open-ended interpretation (which itself gets fixed, if at all, in practice).'

26 Rouse relies here on an argument that *writing as such* is a process of objectification. He quotes Samuel Todes as one source of this argument: 'sensuous characters ... compared to natural shadows are *radically distinct* in their mode of representation. They have an absolute, a universally univocal, sense in place of the contextually variant and analogical sense of natural shadows. This is due to the character of the blank page (or other blank inscribed surface) which serves as a uniform and universal, rather than variegated and local, setting in which sensuous characters can appear' (Todes 1975, 111; quoted in 1987, 78).

27 Walter Vincenti (1990), in *What Engineers Know and How They Know It: Analytical Studies from Aeronautical History,* shows that modern engineering knowledge is autonomous rather than a routine application of theoretical, scientific knowledge. His ideas are discussed in chapter 6.

28 Fleck's small book, first published in 1935 but not brought to bear on science studies until the 1960s, by Thomas Kuhn, is so rich in its pioneering ideas that contemporary social studies of science often read as commentary upon it.

29 Fleck's example of the esoteric communication involved in the stabilization of a particular statement, albeit short-lived and ultimately untenable, one built by consensus and originating in disputation, nonetheless illustrates an important feature of the production of vademecum science: 'the rather impersonal mechanism of its origin, without anybody's deliberate intention or responsibility, can serve as a paradigm for typical principles of vademecum science. Very often it is impossible to find any originator for an

idea generated during discussion and critique. Its meaning changes rapidly; it is adapted and becomes common property. Accordingly it achieves a superindividual value, and becomes an axiom, a guideline for thinking' (1979, 121).

30 In his brief Foreword to Fleck's book, Thomas Kuhn sees the construction of the vademecum as one of Fleck's most important, and unexplored, contributions. He stresses the conscious selections involved in its production, that is, 'the essential and creative act of the individuals who add order and authority by selective systematization within a vademecum' (ix). For Kuhn, a benefit of Fleck's concept is that it allows empirical demonstrations of the production of the vademecum.

31 See Frohmann 1990; 1992.

6. Documenting Universality

1 Recent attention in faculties, schools, and departments of LIS or their renamed successors to 'social informatics,' or investigations of the social context of information production, organization, circulation, and use, is one indicator of interest in issues beyond those posed by epistemological approaches. Another indicator is anecdotal, but I think telling nonetheless. My experience of doctoral students in some leading North American graduate programs suggests a marked degree of frustration in their attempts to find useful research and instruction in LIS about the many and varied cultural effects of the practices in which discursive objects are engaged. The notions of information and communication, with their all too typical reduction of the significance of a doctoral dissertation to increasing the efficiency of an 'information system,' are simply too limited to address such concerns. Nor does a transfer to other disciplines meet their needs. LIS has sensitized students to the importance of specific kinds of documents and documentary systems, which are rarely the primary focus of work elsewhere. Reasonable suspicions of epistemological interpretations of documents suggest satisfying alternatives to research in LIS, in addition to providing a bridge to a literature already too long ignored by many in the field.

2 Harding elaborates: 'For example, Egyptian mystical philosophies and premodern European alchemical traditions were far more useful to the development of sciences in Europe than is suggested by the conventional view that these are only irrational and marginally valuable elements of premodern thought. The Greek legacy of scientific and mathematical thought was not only fortuitously preserved but also developed in Islamic culture, to be claimed by the sciences of the European Renaissance. Some knowledge tra-

ditions that were appropriated and fully integrated into modern sciences are not acknowledged at all. Thus, the principles of pre-Columbian agriculture, that provided potatoes for almost every European ecological niche and thereby had a powerful effect on the nutrition and subsequent history of Europe, was subsumed into European science. Mathematical achievements from India and Arabic cultures provide other such examples. The magnetic needle, rudder, gunpowder, and many other technologies useful to Europeans and the advancement of their sciences (were these not part of scientific instrumentation?) were borrowed from China. Knowledge of local geographies, geologies, animals, plants, classification schemes, medicines, pharmacologies, agriculture, navigational techniques, and local cultures that formed significant parts of European sciences' picture of nature were provided in large part by the knowledge traditions of non-Europeans' (34–5).

3 Lorraine Daston and Katherine Park point out that medieval and early modern fascination with natural wonders and marvels did not conflict with the scientific imagination: 'Medieval and early modern naturalists invoked an order of nature's customs rather than natural laws, defined by marvels as well as by miracles. Although highly ordered, this nature was neither unexceptionably uniform not homogeneous over space and time. Wonders tended to cluster at the margins rather than at the centre of the known world, and they constituted a distinct ontological category, the preternatural, suspended between the mundane and the miraculous. In contrast, the natural order moderns inherited from the late seventeenth and eighteenth centuries is one of uniform, inviolable laws' (1998, 14).

4 Nunberg (1996) shows that our contemporary meaning of 'information' is a nineteenth-century phenomenon.

5 The problem of constructing science based upon singular experiences of unique events has an analogue in the rise of a science of individual persons, as Foucault recognized. In his history of the 'sciences of man,' he makes the connection quite explicitly: 'One is no doubt right to pose the Aristotelian problem: is a science of the individual possible and legitimate?' (1979, 191). And like Dear, Foucault insists on the importance of the constitutive effects of writing: 'These small techniques of notation, of registration, of constituting files, of arranging facts in columns and tables that are so familiar to us now, were of decisive importance in the epistemological "thaw" of the sciences of the individual ... one should look into these procedures of writing and registration, one should look into the mechanisms of examination, into the formation of the mechanisms of discipline, and of a new kind of power over bodies ... The disciplinary methods ... made of this description a

means of control and a method of domination. [The chronicle of a man] is no longer a monument for future memory, but a document for possible use' (190–1).

6 The extent of this limitation is evident in the following observation: 'For practical reasons alone the number of direct witnesses for experimental performances was always limited: in Boyle's laboratory that public probably consisted of at most three to six competent colleagues, and audiences for Royal Society trials rarely exceeded twenty and were typically much smaller' (Shapin 1996, 107).

7 'Boyle himself,' Shapin reports, 'had little if anything to do with the physical manipulation of experimental apparatus' (1996, 394). That job was left to his technicians and assistants, chief among them being Robert Hooke. It was he who performed the 'trials' of experiments in his own home, and the 'shows' in the public rooms of the Royal Society.

8 The differences between contemporary facticity and that of the late seventeenth century, and the reliance of each on specific literary forms, is nicely captured in the closing remarks of Daston's paper: 'Strange facts no longer play a central role in scientific empiricism. They have been replaced by the facts of the replicable, the homogeneous, and the countable. Facts retain their reputation for intractability, but that reputation now rests on different grounds: the first scientific facts were stubborn not because they were robust, resisting all attempts to sweep them under the rug, but rather because they were outlandish, resisting all attempts to subsume them under theory. The literary forms required differed accordingly: descriptions that emphasized differences at the expense of resemblances, that prized rather than ignored peculiarities and variability, that captured surface appearances rather than probed underlying causes. Numbers belong to such descriptions but statistics do not, for statistics assume identical, repeating units. Qualities and quantities can co-exist peacefully in these descriptions, but not generalizations, for strange facts are immiscible' (1998, 38).

9 See Harding 1998, chapters 6 and 7, and Hess 1995, chapters 7 and 8.

10 The reduction of 'human reality to the plane of meaning' is familiar to readers of LIS literature that interprets 'information seeking' as a search for meaning or knowledge. These searches are often described in terms of repairing gaps in an individual's 'picture of the world,' or 'mental model.' See especially Belkin 1980; Belkin 1990.

11 Proposals to protect rock fans from 'improper' messages are a strategy of the Parents' Music Resource Center, founded by, among others, Tipper Gore, the wife of former Vice-President Albert Gore. For a discussion, see Frohmann 2001.

12 On the romance novel, see, for example Radway 1991; on pornography, see Williams 1989, A. Ross 1989, and Hunt 1993.

13 See especially Eisenstein 1979 and Johns 1998.

14 The concept of hybrids or 'quasi-objects' comes from Latour; see especially Latour 1993. For an application to issues arising in LIS, see Frohmann 1995.

15 Grossberg's description makes the point clearly enough: 'The concept of articulation provides a useful starting point for describing the process of forging connections between practices and effects, as well as enabling practices to have different, often unpredicted effects. Articulation is the production of identity on top of difference, of unities out of fragments, of structures across practices. Articulation links this practice to that effect, this text to that meaning, this meaning to that reality, this experience to those politics. And these links are themselves articulated into larger structures' (1992, 54).

16 It would be a mistake to suppose that its incursion is a recent phenomenon. David Noble has charted its role in the development of higher education in America in his *America By Design* (1977). For studies of the amenability of scientific research to corporate and military agendas, see Galison and Hevly 1992 and Greenberg 1999; 2001.

17 The concept of information as a trope originates in the work of Ron Day. See especially his *The Modern Invention of Information: Discourse, History, and Power* (2001).

18 The question of a 'declaration of independence' for documentation was one of the issues addressed in the closing session of DOCAM'03, the first annual meeting of the Document Academy, which is centred upon the concept of the document. Convened by Neils Lund of the the Department of Documentation Science (dokvit) at the University of Tromsø with the assistance of Michael Buckland of the School of Information Management and Systems (SIMS) at the University of California, Berkeley, the conference was held from 13 to 15 August 2003, at SIMS, UC Berkeley. Eleven papers were presented, which are available through: http://thedocumentacademy.hum. uit.no.

7. Documenting Stability

1 The insistence on a *cultural* practice is important, since in other cultural practices, even completing the arithmetical inscription by writing '7' may have no *mathematical* meaning. Wittgenstein makes the point by imagining a culture in which arithmetical expressions are completed by writing the same symbols we do, but in which the point of the writing is to decorate walls (1976, 36, 39–40).

2 See A. Ross 1991a.

3 See Lenoir 1999 for the effects on information technologies on biomedicine.

4 Beniger argues that digitization's transformation of information into a generalized medium for processing and exchange may be as important as the institution of common currencies and exchange rates, which created the world economy. The tendency towards a global system of science creates a crisis of control by threatening the previous barriers of language differences, international politics, national information policies, laws concerning intellectual property, and the incompatibility of information systems. By providing a criterion of the value of scientific contributions, citations function to distribute and communicate status across institutional and organisational lines, unlike local and regional status differentials. Thus citations serve science much as money serves political economy: both are 'symbolic systems that translate status across social contexts' (1988, 24).

5 See especially his comments in 'Reproduction and Difference' in chapter 5 of Rheinberger 1997, 74–83.

6 The role of a support in a discursive formation is central to Paul N. Edwards's analysis of the closed world discourse formed by the intersections of computerization, the imperatives of the military in the Cold War, and cognitive science. He calls a discourse a 'self-elaborating "heterogeneous ensemble" that combines techniques and technologies, metaphors, language, practices, and fragments of other discourses around a support or supports' (1997, 40). In his analysis, the computer functions as a support 'in closed-world and cyborg discourses' (38). A detailed analysis of the function of the library in objectifying discourses has yet to be written.

7 Radford's very insightful reading of the fantasia of the library reaches an unfortunate but familiar technophilic coda in his 'Flaubert, Foucault, and the biblioteque fantastique: toward a postmodern epistemology for library science' (1998, 627). Foucault's careful insistence on the scholarly rigour involved in the emergence of imagination from the intervals created by the 'tight enclosure' of the library's arrangement of titles is lost in Radford's conclusion (one likely quite welcome to corporate sloganeers) that simply by retrieving masses of titles through World Wide Web browsers and Internet search engines, 'every modern library user becomes Flaubert writing *La Tentation*' (630). The 'meticulous erudition,' the 'scholar's patience,' the 'erudite precision,' and the 'untiring attention' that Foucault attributes to the hard work of exploring the intervals between the 'tenets of order' manifested in traditional library classifications are trivialized by reduction to the click of a mouse.

Works Cited

Allen, T.J. 1969. Information needs and uses. In *Annual Review of Information Science and Technology,* vol. 4, ed. C.A. Caudra and A.W. Luke, 3–29. Chicago: Encyclopaedia Britannica.

Anderson, F. H. 1948. *The philosophy of Francis Bacon.* Chicago: University of Chicago Press.

Apostle, R., and B. Raymond. 1997. *Librarianship and the information paradigm.* Lanham, MD: Scarecrow Press.

Bacon, F. 1960. *The New Organon and related writings.* Ed. F.H. Anderson. New York: Liberal Arts Press.

Baker, G.P., and P.M.S. Hacker. 1983. *Wittgenstein: Meaning and understanding.* Essays on the *Philosophical Investigations.* Chicago: University of Chicago Press.

Bazerman, C. 1983. Scientific writing as a social act: A review of the literature of the sociology of science. In *New essays in technical and scientific communication: Research, theory, practice,* ed. P.V. Anderson, R.J. Brockmann, and C.R. Miller, 156–84. Baywood's Technical Communications Series, vol. 2. Farmingale, NY: Baywood Publishing Company.

– 1988a. Physicists reading physics: Schema-laden purposes and purpose-laden schema. In *Shaping written knowledge: The genre and activity of the experimental article in science,* by C. Bazerman, 235–53. Madison: University of Wisconsin Press.

– 1988b. *Shaping written knowledge: The genre and activity of the experimental article in science.* Madison: University of Wisconsin Press.

– 1994. Discourse analysis and social construction. In *Constructing experience,* by C. Bazerman, 114–19. Carbondale and Edwardsville: University of Illinois Press.

Belkin, N.J. 1980. Anomalous states of knowledge as a basis for information retrieval. *Canadian Journal of Information Science* 5:133–43.

– 1990. The cognitive viewpoint in information science. *Journal of Information Science* 16:11–15.

Belkin, N.J., and S.E. Robertson. 1976. Information science and the phenomenon of information. *Journal of the American Society for Information Science* 26:197–204.

Beniger, J.R. 1988. Information society and global science. Telescience: scientific communication in the information age. *Annals of the American Academy of Political and Social Science* 495 (January):14–28.

Bennett, G.E. 1988. *Librarians in search of science and identity: The elusive profession*. Metuchen, NJ: Scarecrow Press.

Bernal, J.D. 1939. *The social function of science*. London: Routledge.

– 1959. The transmission of scientific information: A user's analysis. In *Proceedings of the International Conference on Scientific Information, Washington, D.C., November 16–21, 1958,* 77–95. Washington, DC: National Academy of Sciences – National Research Council.

Bitz, A., A. McAlpine, and R.D. Whitely. 1975. *The production, flow and use of information in research laboratories in different sciences*. Manchester: Manchester Business School and Centre for Business Research.

Bloor, D. 1976. *Knowledge and social imagery*. London: Routledge and Kegan Paul.

Bourdieu, P. 1988. *Homo academicus*. Trans. P. Collier. Stanford: Stanford University Press.

– 1991. *Language and symbolic power*. Ed. J.B. Thompson. Trans. G. Raymond and M. Adamson. Cambridge: Harvard University Press.

Brookes, B.C. 1980a. The foundations of information science. Part I. Philosophical aspects. *Journal of Information Science* 2:125–33.

– 1980b. The foundations of information science. Part II. Quantitative aspects: Classes of things and the challenge of human individuality. *Journal of Information Science* 2:209–21.

– 1980c. The foundations of information science. Part III. Quantitative aspects: Objective maps and subjective landscapes. *Journal of Information Science* 2:269–75.

– 1981. The foundations of information science. Part IV. Information science: The changing paradigm. *Journal of Information Science* 3:3–12.

Brown, J.S., and P. Duguid. 2000. *The social life of information*. Boston: Harvard Business School Press.

Buckland, M. 1996. Documentation, information science, and library science in the U.S.A. *Information Processing and Management* 32(1):63–75.

– 1997. What is a document? *Journal of the American Society for Information Science* 48(9):804–9.

Bush, V. 1991. As we may think. In *From memex to hypertext: Vannevar Bush and the mind's machine,* ed. J.M. Nyce and P. Kahn, 85–107. Boston: Academic Press.

Callon, M. 1986. Some elements of a sociology of translation: Domestication of the scallops and the fishermen of St. Brieuc Bay. In *Power, action and belief: A new sociology of knowledge?* ed. J. Law. Sociological Review Monograph, no. 32, 196–233. London: Routledge and Kegan Paul.

– 1990. Society in the making: The study of technology as a tool for sociological analysis. In *The social construction of technological systems: New directions in the sociology and history of technology,* ed. W.E. Bijker, T.P. Hughes, and T.J. Pinch, 81–103. Cambridge, MA: MIT Press.

Callon, M., J. Law, and A. Rip. 1986. *Mapping the dynamics of science and technology: Sociology of science in the real world.* London: Macmillan Press.

Case, D.O. 2002. *Looking for information: A survey of research on information seeking, needs, and behavior.* New York: Academic Press.

Clanchy, M.T. 1993. *From memory to written record, England 1066–1307.* Oxford: Blackwell.

Collins, H.M. 1992. *Changing order: Replication and induction in scientific practice.* Chicago: University of Chicago Press.

Crane, D. 1971. Information needs and uses. In *Annual Review of Information Science and Technology,* vol. 6, ed. C.A. Caudra and A.W. Luke, 3–39. Chicago: Encyclopaedia Brittanica.

Crawford, S. 1978. Information needs and uses. In *Annual Review of Information Science and Technology,* vol. 13, ed. M.E. Williams, 61–81. New York: Knowlege Industry Publications.

Crawford, S.Y., J.M. Hurd, and A.C. Weller. 1996. *From print to electronic: The transformation of scientific communication.* Medford, NJ: Information Today.

Crombie, A.C. 1994. *Styles of scientific thinking in the European tradition: The history of argument and explanation in the mathematical and biomedical sciences and arts.* London: Duckworth.

– 1996. *Science, art and nature in medieval and modern thought.* London: Hambledon Press.

Cronin, B. 1984. *The citation process: The role and significance of citations in scientific communication.* London: Taylor Graham.

Cronin, B., and K. Overfelt. 1994. Citation-based auditing of academic performance: Citation analysis of faculty authors at Indiana University School of Library and Information Science, 1982–1993. *Journal of the American Society for Information Science* 45:62–71.

Daston, L. 1998. The language of strange facts in early modern science. In *Inscribing science: Scientific texts and the materiality of communication*, ed. T. Lenoir. Writing Science. Gen. eds. T. Lenoir and H.U. Gumbrecht, 20–38. Stanford: Stanford University Press.

Daston, L., and K. Park. 1998. *Wonders and the order of nature, 1150–1750*. New York: Zone Books.

Davies, R. 1989. The creation of new knowledge by information retrieval and classification. *Journal of Documentation* 45(4):273–301.

Davis, E. 1998. *Techgnosis: Myth, magic and mysticism in the age of information*. New York: Three Rivers Press.

Day, R. 1997. Paul Otlet's book and the writing of social space. *Journal of the American Society for Information Science* 48: 310–17.

– 2001. *The modern invention of information: Discourse, history, and power*. Carbondale and Edwardsville: Southern Illinois University Press.

Dear, P. 1991. Narratives, anecdotes, and experiments: Turning experience into science in the seventeenth century. In *The literary structure of scientific argument: Historical studies*, ed. P. Dear, 135–63. Philadelphia: University of Pennsylvania Press.

– 1995. *Discipline and experience: The mathematical way in the scientific revolution*. Chicago: University of Chicago Press.

Debons, A., ed. 1974. *Information science: Search for identity. Proceedings of the 1972 NATO Advanced Study Institute in Information Science*. Books in library and information science, vol. 7. New York: Marcel Dekker.

Dervin, B., and M. Nilan. 1986. Information needs and uses. *Annual Review of Information Science and Technology*, vol. 21, ed. M.E. Williams, 3–33. New York: Knowledge Industry.

Edwards, P.N. 1997. *The closed world: Computers and the politics of discourse in Cold War America*. Cambridge, MA: MIT Press.

Egan, M., and H.H. Henkle. 1956. Ways and means in which research workers, executives, and others use information. In *Documentation in action*, ed. J.H. Shera, A. Kent, and J.W. Perry, 137–59. New York and London: Reinhold Publishing Corporation; Chapman & Hall.

Eisenstein, E. 1979. *The printing press as an agent of change: Communication and cultural transformations in early modern Europe*. Cambridge: Cambridge University Press.

Ellis, D. 1989. A behavioural approach to information retrieval system design. *Journal of Documentation* 45(3):171–212.

– 1993. Modeling the information-seeking patterns of academic researchers: A grounded theory approach. *Library Quarterly* 63(4):469–86.

Ellis, D., D. Cox, and K. Hall. 1993. A comparison of the information seeking patterns of researchers in the physical and social sciences. *Journal of Documentation* 49(4):356–69.

Erdelez, S. 1997. Information encountering: A conceptual framework for accidental information discovery. In *Information seeking in context: Proceedings of an International Conference on Research in Information Needs, Seeking, and Uses in Different Contexts, 14–16 August, Tampere, Finland*, ed. P. Vakkari, R. Savolainen, and B. Dervin, 412–20. London: Taylor Graham.

Fleck, L. 1979. *Genesis and development of a scientific fact*. Trans. F. Bradley and T.J. Trenn. Chicago: University of Chicago Press.

Foucault, M. 1972. *The archaeology of knowledge and the discourse on language.* Trans. A.M.S. Smith. New York: Harper & Row, Harper Colophon.

– 1977. Fantasia of the library. In *Language, counter-memory, practice: Selected essays and interviews*, ed. D.F. Bouchard, trans. S. Simon and D.F. Bouchard, 87–109. Ithaca: Cornell University Press.

– 1979. *Discipline and punish: The birth of the prison*. Trans. A. Sheridan. New York: Random House, Vintage.

Frasca-Spada, M., and N. Jardine, eds. 2000. *Books and the sciences in history.* Cambridge: Cambridge University Press.

Frohmann, B. 1990. Rules of indexing: A critique of mentalism in information retrieval theory. *Journal of Documentation* 46(2):81–101.

– 1992. The power of images: A discourse analysis of the cognitive viewpoint. *Journal of Documentation* 48(4):365–86.

– 1995. Taking information policy beyond information science: Applying the actor network theory. In *Connectedness: Information, systems, people, organizations. Proceedings of the 23rd Annual Conference of the Canadian Association for Information Science, 7–10 June 1995, Edmonton, Alberta*, ed. H.A. Olson and D.B. Ward, 19–28. Edmonton: School of Library and Information Studies, University of Alberta.

– 1998. Review of Librarianship and the Information Paradigm, by Richard Apostle and Boris Raymond. Lanham, MD: Scarecrow Press, 1997. *Library Quarterly* 68:498–500.

– 2001. Discourse and documentation: Some implications for pedagogy and research. *Journal of Education for Library and Information Science* 42(1):13–26.

Fuller, S. 1992. Social epistemology and the research agenda of science studies. In *Science as practice and culture*, ed. A. Pickering, 390–428. Chicago: University of Chicago Press.

– 1995a. Cybermaterialism, or why there is no free lunch in cyberspace. *The Information Society* 11:325–32.

– 1995b. Cyberplatonism: An inadequate constitution for the republic of science. *The Information Society* 11:293–303.

Galison, P. 1987. *How experiments end*. Chicago: University of Chicago Press.
– 1997. *Image and logic: A material culture of microphysics*. Chicago: University of Chicago Press.
– 1999. Trading zone: Coordinating action and belief. In *The science studies reader*, ed. M. Biagioli, 137–60. New York: Routledge.
– 2003. *Einstein's clocks, Poincaré's maps: Empires of time*. New York and London: W.W. Norton.
Galison, P., and B. Hevly, eds. 1992. *Big science: The growth of large-scale research*. Stanford: Stanford University Press.
Galison, P., and D.J. Stump, eds. 1996. *The disunity of science: Boundaries, contexts, power*. Writing Science. Gen. eds. T. Lenoir and H.U. Gumbrecht. Stanford: Stanford University Press.
Garfield, E. 1979. *Citation indexing: Its theory and application in science, technology and humanities*. New York: Wiley.
– 1981a. ABCs of cluster mapping. Part 1. Most active fields in the life sciences in 1978. In Garfield, *Essays of an information scientist*, 634–41. Philadelphia: ISI Press.
– 1981b. ABCs of cluster mapping. Part 2. Most active fields in the physical sciences in 1978. In Garfield, *Essays of an information scientist*, 642–49. Philadelphia: ISI Press.
– 1981–2. Introducing ISI/GeoSciTech and the GeoSciTech Citation Index * the 50 most-active research fronts in 1981 in the earth sciences illustrate the unique retrieval capabilities of our new online and print services. In Garfield, *Essays of an information scientist*, 607–14. Philadelphia: ISI Press.
Garfield, E., M.V. Malin, and H. Small. 1978. Citation data as science indicators. In *Toward a metric of science: The advent of science indicators*, ed. Y. Elkana, et al., 179–207. New York: John Wiley & Sons.
Garvey, W.D. 1979. *Communication: The essence of science. Facilitating information exchange among librarians, scientists, engineers and students*. Oxford, New York, and Toronto: Pergamon Press.
Gilbert, G.N., and M. Mulkay. 1984. *Opening Pandora's box: A sociological analysis of scientists' discourse*. Cambridge: Cambridge University Press.
Greenberg, D.S. 1999. *The politics of pure science*. New ed. Chicago: University of Chicago Press.
– 2001. *Science, money, and politics: Political triumph and ethical erosion*. Chicago: University of Chicago Press.
Griffith, B.C., H.G. Small, J.A. Stonehill, and S. Dey. 1974. The Structure of scientific literatures II: Toward a macro- and microstructure for science. *Science Studies* 4(4):339–65.

Grossberg, L. 1992. *We gotta get out of this place: Popular conservatism and post-modern culture*. New York: Routledge.

Gumbrecht, H.U., and K.L. Pfeiffer. 1994. *Materialities of communication*. Trans. W. Whobrey. Writing Science. Gen. eds. T. Lenoir and H.U. Gumbrecht. Stanford: Stanford University Press.

Hacking, I. 1983. *Representing and intervening: Introductory topics in the philosophy of natural science*. Cambridge: Cambridge University Press.

– 1992. The self-vindication of the laboratory sciences. In *Science as practice and culture*, ed. A. Pickering, 29–64. Chicago: University of Chicago Press.

– 1996. The disunities of the sciences. In *The disunity of science: Boundaries, contexts, and power*, ed. P. Galison and D.J. Stump. Writing Science. Gen. eds. T. Lenoir and H.U. Gumbrecht, 37–74. Stanford: Stanford University Press.

Hagstrom, W.O. 1965. *The scientific community*. New York: Basic Books.

Haraway, D.J. 1996. Modest witness: Feminist diffractions in science studies. In *The disunity of science: Boundaries, contexts, power*, ed. P. Galison and D.J. Stump. Writing Science. Gen. eds. T. Lenoir and H.U. Gumbrecht, 428–41. Stanford: Stanford University Press.

Harding, S. 1998. *Is science multicultural? Postcolonialisms, feminisms, and epistemologies*. Bloomington and Indianapolis: Indiana University Press.

Harnad, S. 1995a. The PostGutenberg galaxy: How to get there from here. *The Information Society* 11:285–91.

– 1995b. Sorting the esoterica from the exoterica: There's plenty of room in cyberspace. A response to Fuller. *The Information Society* 11:305–24.

Harris, R. 1992. *Librarianship: The erosion of a woman's profession*. Norwood, NJ: Ablex.

Hayes, R.M. 1992. The needs of science and technology. *Science and Technology Libraries* 12(4):3–33.

Hayles, N.K. 1999. *How we became posthuman: Virtual bodies in cybernetics, literature, and informatics*. Chicago: University of Chicago Press.

Herner, S., and M. Herner. 1967. Information needs and uses in science and technology. In *Annual Review of Information Science and Technology*, vol. 2, ed. C.A. Caudra, 1–34. New York: Interscience Publishers.

Hernon, P., and C. Schwartz. 1993. Library and information science research: Is it misunderstood? *Library and Information Science Research* 15(3):215–17.

Hess, D.J. 1995. *Science and technology in a multicultural world: The cultural politics of facts and artifacts*. New York: Columbia University Press.

Hughes, T.P. 1998. *Rescuing Prometheus*. New York: Pantheon Books.

Hunt, L., ed. 1993. *The invention of pornography: Obscenity and the origins of modernity, 1500–1800*. New York: Zone Books.

Johns, A. 1998. *The nature of the book: Print and knowledge in the making*. Chicago: University of Chicago Press.

Jones, C.A., and P. Galison, eds. 1998. *Picturing science, producing art*. In collaboration with A. Slaton. New York: Routledge.

Keresztesi, M. 1982. The science of bibliography: Theoretical implications for bibliographic instruction. In *Theories of bibliographic education: Designs for teaching*, ed. C. Oberman and K. Strauch, 1–26. New York: R.R. Bowker.

Knorr, K.D. 1979. Tinkering toward success: Prelude to a theory of scientific practice. *Theory and Society* 8:347–76.

Knorr Cetina, K. 1999. *Epistemic cultures: How the sciences make knowledge*. Cambridge: Harvard University Press.

Knorr-Cetina, K.D. 1981. *The manufacture of knowledge: An essay on the constructivist and contextual nature of science*. Oxford: Pergamon Press.

Latour, B. 1983. Give me a laboratory and I will raise the world. In *Science observed: Perspectives on the social study of science*, ed. K.D. Knorr-Cetina and M. Mulkay, 141–70. London: Sage Publications.

– 1987. *Science in action: How to follow scientists and engineers through society*. Cambridge: Harvard University Press.

– 1988. *The Pasteurization of France*. Cambridge: Harvard University Press.

– 1992. Where are the missing masses? The sociology of a few mundane artifacts. In *Shaping technology/building society: Studies in sociotechnical change*, ed. W.E. Bijker and J. Law, 225–58. Cambridge: MIT Press.

– 1993. *We have never been modern*. Trans. C. Porter. Cambridge: Harvard University Press.

– 1996. *Aramis or the love of technology*. Trans. C. Porter. Cambridge: Harvard University Press.

– 1999. *Pandora's hope: Essays on the reality of science studies*. Cambridge: Harvard University Press.

Latour, B., and S. Woolgar. 1986. *Laboratory life: The construction of scientific facts*. 2nd ed. Introd. by J. Salk. Princeton: Princeton University Press.

Law, J. 1990. Technology and heterogeneous engineering: The case of Portuguese expansion. In *The social construction of technological systems: New directions in the sociology and history of technology*, ed. W.E. Bijker, T.P. Hughes, and T.J. Pinch, 111–34. Cambridge, MA: MIT Press.

Lenoir, T., ed. 1998. *Inscribing science: Scientific texts and the materiality of communication*. Writing Science. Gen. eds. T. Lenoir and H.U. Gumbrecht. Stanford: Stanford University Press.

– 1997. *Instituting science: The cultural production of scientific disciplines*. Writing Science. Gen. eds. T. Lenoir and H.U. Gumbrecht. Stanford: Stanford University Press.

– 1999. Shaping biomedicine as an information science. In *Proceedings of the 1998 Conference on the History and Heritage of Science Information Systems*, ed. M.E. Bowden, T.B. Hahn, and R.V. Williams, 27–45. ASIS Monograph Series. Medford, NJ: Information Today, Inc.

Lévy, P. 1997. *Collective intelligence: Mankind's emerging world in cyberspace.* Trans. R. Bononno. New York: Plenum Trade.

Lin, N., and W.D. Garvey. 1972. Information needs and uses. In *Annual Review of Information Science and Technology*, vol. 7, ed. C.A. Caudra and A.W. Luke, 3–37. Washington, DC: American Association for Information Science.

Lipetz, B.-A. 1970. Information needs and uses. In *Annual Review of Information Science and Technology*, vol. 5, ed. C.A. Caudra and A.W. Luke, 3–32. Chicago: Encyclopaedia Brittannica.

Lord, A.B. 1960. *The singer of tales.* Harvard studies in comparative literature, no. 24. Cambridge: Harvard University Press.

Lynch, M. 1985. *Art and artifact in laboratory science: A study of shop work and shop talk in a research laboratory.* Studies in Ethnomethodology. Ed. H. Garfinkel. London: Routledge and Kegan Paul.

Lynch, M., and S. Woolgar, eds. 1990. *Representation in scientific practice.* Cambridge, MA: MIT Press.

Martin, J. 1992. *Francis Bacon, the state, and the reform of natural philosophy.* Cambridge: Cambridge University Press.

Martyn, J. 1974. Information needs and uses. In *Annual Review of Information Science and Technology*, vol. 9, ed. C.A. Caudra, A.W. Luke, and J.L. Harris, 3–23. Washington, DC: American Association for Information Science.

McKenzie, P.J. 2003. A model of information practices in accounts of everyday-life information seeking. *Journal of Documentation* 59(1):19–40.

Medawar, P. 1990. Is the scientific paper a fraud? In *The threat and the glory*, ed. D. Pyke, foreword by L. Thomas, 228–33. New York: HarperCollins.

Menzel, H. 1964. The information needs of current scientific research. *Library Quarterly* 34:4–19.

– 1966. Information needs and uses in science and technology. In *Annual Review of Information Science and Technology*, vol. 1, ed. C.A. Caudra. American Documentation Institute Annual Review, 41–69. New York: John Wiley & Sons.

Merton, R.K. 1973a. Behaviour patterns of scientists. In *The sociology of science: Theoretical and empirical investigations*, ed. N.W. Storer, 323–42. Chicago: University of Chicago Press.

– 1973b. The Matthew effect in science. In *The sociology of science: Theoretical and empirical investigations*, ed. N.W. Storer, 439–59. Chicago: University of Chicago Press.

- 1973c. Multiple discoveries as strategic research site. In *The sociology of science: Theoretical and empirical investigations*, ed. N.W. Storer, 371–82. Chicago: University of Chicago Press.
- 1973d. The normative structure of science. In *The sociology of science: Theoretical and empirical investigations*, ed. N.W. Storer, 267–78. Chicago: University of Chicago Press.
- 1973e. Priorities in scientific discovery. In *The sociology of science: Theoretical and empirical investigations*, ed. N.W. Storer, 287–324. Chicago: University of Chicago Press.

Merton, R.K., and H. Zuckerman. 1973. Institutionalized patterns of evaluation in science. In *The sociology of science: Theoretical and empirical investigations*, ed. N.W. Storer, 461–96. Chicago: University of Chicago Press.

Mikhailov, A.I., A.I. Chernyi, and R.S. Giliarevski. 1984. *Scientific communications and informatics*. Trans. R.H. Burger. Academy of Sciences, State Committee on Science and Technology (VINITI). Arlington, VA: Information Resources Press.

Miksa, F.L. 1998. *The DDC, the universe of knowledge, and the post-modern library*. Albany, NY: Forest Press.

Mulkay, M.J., and A.T. Williams. 1971. A sociological study of a physics department. *British Journal of Sociology* 22:68–82.

Neill, S.D. 1982. Brookes, Popper, and objective knowledge. *Journal of Information Science* 4(1):33–9.
- 1987. The dilemma of the subjective in information organization and retrieval. *Journal of Documentation* 43(3):193–211.

Noble, D.F. 1977. *America by design: Science, technology, and the rise of corporate capitalism*. New York: Knopf.

Nunberg, G. 1996. Farewell to the information age. In *The future of the book*, ed. G. Nunberg, 103–38. Berkeley: University of California Press.

Nyce, J.M., and P. Kahn, eds. 1991. *From Memex to hypertext: Vannevar Bush and the mind's machine*. Boston: Academic Press.

Otlet, P. 1990. *International organisation and dissemination of knowledge: Selected essays of Paul Otlet*. Ed. and trans. B. Rayward. FID, no. 684. Amsterdam: Elsevier.

Paisley, W.J. 1968. Information needs and uses. In *Annual Review of Information Science and Technology*, vol. 3, ed. C.A. Caudra, 1–30. Chicago: Encyclopaedia Britannica.

Pear, R. 1999. N.I.H. plan for journal on the Web draws fire. *New York Times*, 8 June.

Pettigrew, K.E. 1999. Waiting for chiropody: Contextual results from an ethnographic study of the information behaviour among attendees at community clinics. *Information Processing and Management* 35:801–17.

Pickering, A., ed. 1992. *Science as practice and culture.* Chicago: University of Chicago Press.

– 1995. *The mangle of practice: Time, agency, and science.* Chicago: University of Chicago Press.

Pinelli, T.E. 1991. The information-seeking habits and practices of engineers. In *Information seeking and communicating behavior of scientists and engineers,* ed. C. Steinke, 5–25. New York: Hayworth Press.

Popper, K.R. 1965. *Conjectures and refutations: The growth of scientific knowledge.* 2nd ed. New York: Basic.

– 1968. *The logic of scientific discovery.* New York: Harper and Row, Harper Torchbooks.

– 1972. *Objective knowledge: An evolutionary approach.* Oxford: Clarendon Press.

Poster, M. 1996. Databases as discourse, or, electronic interpellations. In *Computers, surveillance, and privacy,* ed. D. Lyon and E. Zureik, 175–92. Minneapolis: University of Minnesota Press.

Price, D.J.d.S. 1970. Citation measures of hard science, soft science, technology, and nonscience. In *Communication among scientists and engineers,* ed. C.E. Nelson and D. Pollack, 3–22. Lexington, MA: Lexington Books, D.C. Heath and Company.

– 1986. *Little science, big science ... and beyond.* New York: Columbia University Press.

Radford, G.P. 1992. Positivism, Foucault, and the fantasia of the library: Conceptions of knowledge and the modern library experience. *Library Quarterly* 62(4):408–24.

– 1998. Flaubert, Foucault, and the biblioteque fantastique: Toward a postmodern epistemology for library science. *Library Trends* 46(4):616–34.

Radway, J.A. 1991. *Reading the romance: Women, patriarchy, and popular literature.* Rev. ed. Chapel Hill: University of North Carolina Press.

Ravetz, J.R. 1971. *Scientific knowledge and its social problems.* Oxford: Clarendon Press.

Rayward, W.B. 1967. The UDC and the FID – a historical perspective. *Library Quarterly* 37(3):259–78.

– 1975. *The universe of information: The work of Paul Otlet for documentation and international organisation.* FID Publication 520. Moscow: All-Union Institute for Scientific and Technical Information (VINITI).

– 1991. The case of Paul Otlet, pioneer of information science, internationalist, visionary: Reflections on biography. *Journal of Librarianship and Information Science* 23(3):135–45.

– 1994a. Some schemes for restructuring and mobilising information in documents: A historical perspective. *Information Processing and Management* 30(2):163–75.

– 1994b. Visions of Xanadu: Paul Otlet (1866–1944) and hypertext. *Journal of the American Society for Information Science* 45(4):235–50.

– 1997. The origins of information science and the work of the International Institute of Bibliography/International Federation for Documentation and Information (FID). *Journal of the American Society for Information Science* 48:289–300.

Rheinberger, H.-J. 1997. *Toward a history of epistemic things: Synthesizing proteins in the test tube.* Writing Science. Gen. eds. T. Lenoir and H.U. Gumbrecht. Stanford: Stanford University Press.

– 1998. Experimental systems, graphemic spaces. In *Inscribing science: Scientific texts and the materiality of communication,* ed. T. Lenoir. Writing Science. Gen. eds. T. Lenoir and H.U. Gumbrecht, 285–303. Stanford: Stanford University Press.

Rieusset-Lemarié, I. 1997. P. Otlet's mundaneum and the international perspective in the history of documentation and information science. *Journal of the American Society for Information Science* 48(4):301–9.

Rorty, R. 1979. *Philosophy and the mirror of nature.* Princeton: Princeton University Press.

Ross, A. 1989. *No respect: Intellectuals and popular culture.* New York: Routledge.

– 1991a. New age – a kinder, gentler science? In *Strange weather: Culture, science, and technology in the age of limits,* 1–74. London: Verso.

– 1991b. *Strange weather: Culture, science, and technology in the age of limits.* London: Verso.

– 1998. Jobs in cyberspace. In *Real love: In pursuit of cultural justice,* 7–34. New York: New York University Press.

Ross, C.S. 1999. Finding without seeking: The information encounter in the context of reading for pleasure. *Information Processing and Management* 35:783–99.

Rotman, B. 2000. *Mathematics as sign: Writing, imagining, counting.* Stanford: Stanford University Press.

Rouse, J. 1987. *Knowledge and power: Toward a political philosophy of science.* Ithaca: Cornell University Press.

– 1996. *Engaging science: How to understand its practices philosophically.* Ithaca: Cornell University Press.

Royal Society (Great Britain). 1948. *The Royal Society Scientific Information Con-*

ference, 21 June - 2 July 1948: Report and papers submitted. London: Royal Society.

Savolainen, R. 1995. Everyday life information seeking: Approaching information seeking in the context of 'way of life.' *Library and Information Science Research* 17:259–94.

Schaffer, S. 1998. The Leviathan of Parsonstown: Literary technology and scientific representation. In *Inscribing science: Scientific texts and the materiality of communication*, ed. T. Lenoir. Writing Science. Gen. eds. T. Lenoir and H.U. Gumbrecht, 182–222. Stanford: Stanford University Press.

Schrader, A.M. 1984a. In search of a definition of library and information science. *Canadian Journal of Information Science* 9:59–78.

– 1984b. In search of a name: Information science and its conceptual antecedents. *Library and Information Science Research* 6(3):227–71.

– 1986. The domain of information science: Problems in conceptualisation and in consensus-building. *Information Services and Use* 6(5–6):169–205.

Shapin, S. 1984. Pump and circumstance: Robert Boyle's literary technology. *Social Studies of Science* 14:481–520.

– 1988a. Following scientists around. *Social Studies of Science* 18:533–50.

– 1988b. The house of experiment in seventeenth-century England. *Isis* 79:373–404.

– 1994. *A social history of truth: Civility and science in sixteenth-century England.* Chicago: University of Chicago Press.

– 1996. *The scientific revolution.* Chicago: University of Chicago Press.

Shapin, S., and S. Schaffer. 1985. *Leviathan and the air-pump: Hobbes, Boyle, and the experimental life.* Princeton: Princeton University Press.

Small, H. 1973. Co-citation in the scientific literature: A new measure of the relationship between two documents. *Journal of the American Society for Information Science* 24(4):265–69.

Small, H., and B.C. Griffith. 1974. The structure of scientific literatures I: Identifying and graphing specialities. *Science Studies* 4(1):17–40.

Star, S.L., and J. Griesemer. 1989. Institutional ecology, 'translations,' and boundary objects: Amateurs and professionals in Berkeley's Museum of Vertebrate Zoology, 1907–1939. *Social Studies of Science* 19:387–420.

Stoan, S.K. 1984. Research and library skills: An analysis and interpretation. *College & Research Libraries* 45 (March):99–109.

Süskind, P. 1986. *Perfume: The story of a murderer.* Trans. J.E. Woods. New York: Pocket Books.

Swanson, D.R. 1986. Undiscovered public knowledge. *Library Quarterly* 56(2):103–18.

– 1987. Two medical literatures that are logically but not bibliographically connected. *Journal of the American Society for Information Science* 38(4):228–33.

– 1989a. Online search for logically-related noninteractive medical literatures: A systematic trial-and-error strategy. *Journal of the American Society for Information Science* 40(5):356–58.

– 1989b. A second example of mutually isolated medical literatures related by implicit, unnoticed connections. *Journal of the American Society for Information Science* 40(6):432–35.

– 1990a. The absence of co-citation as a clue to undiscovered causal connections. In *Scholarly communication and bibliometrics*, ed. C.L. Borgman, 129–37. Newbury Park, CA: Sage.

– 1990b. Medical literature as a potential source of new knowledge. *Bulletin of the Medical Library Association* 78(1):29–37.

Todes, S. 1975. Shadows in knowledge. In *Dialogues in phenomenology*, ed. D. Idhe and R. Zaner, 94–113. The Hague: Martinus Nijhoff.

Toren, N. 1983. The scientific ethos debate: A meta-theoretical view. *Social Science and Medicine* 17(21):1665–72.

Traweek, S. 1988. *Beamtimes and lifetimes: The world of high energy physicists.* Cambridge: Harvard University Press.

Tuominen, K., and R. Savolainen. 1997. A social constructivist approach to the study of information use as discursive action. In *Information seeking in context: Proceedings of an international conference on research in information needs, seeking, and use in different contexts, 14–16 August, 1996, Tampere, Finland*, ed. P. Vakkari, R. Savolainen, and B. Dervin, 81–96. London: Taylor Graham.

Vincenti, W.G. 1990. *What engineers know and how they know it: Analytical studies from aeronautical history.* Baltimore: Johns Hopkins University Press.

Weiss, M.J. 1988. *The clustering of America.* New York: Harper and Row.

Wertheim, M. 1999. *The pearly gates of cyberspace: A history of space from Dante to the internet.* New York: W.W. Norton.

Williams, L. 1989. *Hard core: Power, pleasure, and the 'frenzy of the visible.'* Berkeley: University of California Press.

Wilson, E.O. 1998. *Consilience: The unity of knowledge.* New York: Knopf.

Wilson, P. 1977. *Public knowledge, private ignorance: Toward a library and information policy.* Westport, CT: Greenwood Press.

Wilson, T.D. 1984. The cognitive approach to information-seeking behaviour and information use. *Social Science Information Studies* 4:197–204.

– 1994. Information needs and uses: Fifty years of progress? In *Fifty years of information progress: A Journal of Documentation review*, ed. B.C. Vickery, 15–51. London: Aslib, The Association for Information Management.

– 1999. Models in information behaviour research. *Journal of Documentation* 55(3):249–70.

Wittgenstein, L. 1958. *Philosophical investigations*. Trans. G.E.M. Anscombe. New York: Macmillan.

– 1969. *Preliminary studies for the 'Philosophical Investigations,' generally known as the Blue and Brown books*. Oxford: Basil Blackwell.

– 1976. *Wittgenstein's lectures on the foundations of mathematics, Cambridge, 1939; from the notes of R.G. Bosanquet, Norman Malcolm, Rush Rhees, and Yorick Smythies*. Ed. C. Diamond. Ithaca, NY: Cornell University Press.

– 1978. *Philosophical grammar*. Ed. R. Rhees. Trans. A. Kenny. Berkeley: University of California Press.

– 1979. *Wittgenstein's lectures, Cambridge, 1932–1935*. Ed. A. Ambrose. Oxford: Blackwell.

– 1980. *Culture and value*. Ed. G.H. von Wright. In collaboration with H. Nyman, trans. P. Winch. Chicago: University of Chicago Press.

Wurman, R.S. 1989. *Information Anxiety*. New York: Doubleday.

Zachary, G.P. 1997. *Endless frontier: Vannevar Bush, engineer of the American Century*. New York: Free Press.

Zagorin, P. 1998. *Francis Bacon*. Princeton: Princeton University Press.

Index